MEDIA AND COMMUNICATIONS –
TECHNOLOGIES, POLICIES AND CHALLENGES

OPTICAL COMMUNICATION SYSTEMS

FUNDAMENTALS, TECHNIQUES AND APPLICATIONS

MEDIA AND COMMUNICATIONS – TECHNOLOGIES, POLICIES AND CHALLENGES

Additional books in this series can be found on Nova's website under the Series tab.

Additional e-books in this series can be found on Nova's website under the e-book tab.

MEDIA AND COMMUNICATIONS –
TECHNOLOGIES, POLICIES AND CHALLENGES

OPTICAL COMMUNICATION SYSTEMS

FUNDAMENTALS, TECHNIQUES AND APPLICATIONS

KARLA P. BOYD
EDITOR

nova publishers
New York

Library of Congress Cataloging-in-Publication Data

ISBN: 978-1-63482-833-8

Published by Nova Science Publishers, Inc. † New York

CONTENTS

PREFACE

Optical communication, also known as optical telecommunication, is communication at a distance using light to carry information. It can be performed visually or by using electronic devices. In this book, the authors propose systems of microring resonators (MRRs) that use localized spatial and temporal soliton pulses to form optical communication signals. The main advantage of using soliton pulses in optical communication systems is that the shape of the pulse remain almost unaltered over a long distance. The third chapter introduces digital signal processing for Nyquist-WDM systems. The last chapter in this book provides an in-depth analysis performed on the transmission, reflection and directive properties of nanoantennas.

Chapter 1 - A spectrum of light over a broad range can be generated, thus solitons can be utilized to generate chaotic filter features when spreading within nonlinear microring resonators (MRRs). The authors propose a system of MRRs that uses localized soliton pulses to form the optical communication signals. This method uses nonlinear behaviours of light applicable for soliton transmission. Optical carrier generation is the basic building block for communication systems in the optical domain. One important aspect of the MRR system is that tuning of the system parameters allows for desired soliton carriers with specific key characteristics. The normalized output of the light field is defined by the ratio between the output and input fields. The results demonstrate optical soliton pulses localized within the MRRs system with 20,000 roundtrips. A frequency soliton pulse can be formed and trapped within the panda ring resonator system with suitable ring parameters. The throughput output (E_{th}) shows localised soliton pulses with full width at half maximum (FWHM) and free spectral range (FSR) of 5 MHz and 2 GHz, respectively, where soliton pulses at frequencies of 50 and 52 GHz are generated. The drop output (E_d) shows pulses with FWHM and FSR of 10 MHz 2 GHz. The efficiency of this system can be evaluated by the output solitonic carrier signal generated by the MRRs.

Chapter 2 - A system of bright soliton array (BSA) for optical communication using the wired/wireless link is presented. Gaussian laser beam with relevant parameters is input into the ring resonator system. Here, bright soliton with Full Width at Half Maximum (FWHM) and Free Specrum Range (FSR) of 120 ps and 370 ps are generated respectively. BSA can be obtained by a series of microring resonators (MRRs) with the input optical laser pulse. When a soliton pulse is propagated within a nonlinear microring device, a balance should be achieved between the dispersion length (LD) and the nonlinear length (NL), thus, the MRR in the particular case is very similar to an Fabry-Perot cavity, which has an input and output mirror with a field reflectivity. Beside improvements in efficiency and beam quality these soliton sources provide short and ultra-short pulses, leading to improved process efficiencies and new fields of optical soliton communications. In this study, transmission of BSA via 50 km fiber optics and 25 meters wireless has been investigated.

Chapter 3 - In optical fiber transmission systems, there has been an increasing interest in using digital signal processing (DSP) combined with coherent detection to increase the transmission distance, spectral efficiency (SE), etc. Nyquist pulse shaping can be used to improve the spectral efficiency (SE) by narrowing the signal spectrum without introducing inter-symbol interference in a wavelength-division-multiplexed (WDM) system. Such systems are usually called Nyquist-WDM systems. In this chapter, the authors introduce digital signal processing for Nyquist-WDM systems. The author's discussion covers the following topics. (1) Introduction of Nyquist spectral shaping in optical fiber WDM systems. (2) Theoretical background of Digital Nyquist spectral shaping. (3) Transmitter-side DSP and digital-to-analog converter (DAC) for Nyquist spectral shaping. (4) Receiver-side Matched filter design for Nyquist-WDM systems.

Topic (1) is an introduction of Nyquist spectral shaping in optical fiber WDM systems.

Topic (2) introduces theoretical background about Nyquist spectral shaping and digital filters.

Topic (3) discusses about transmitter-side DSP and DAC. In topic (3), the required tap number or overlap length are investigated for finite-impulse-response (FIR) or frequency-domain-equalization (FDE) filters. A FDE filter with a fast Fourier transform (FFT) size of 64 and an overlap length of 8 points performs slightly better than a 17-tap FIR filter. 25 taps are enough for a FIR filter to do Nyquist spectral shaping for PDM-QPSK WDM systems, with negligible optical signal-to-noise ratio (OSNR) penalty. And 37 taps are sufficient for a digital FIR filter to generate Nyquist-WDM PDM-16QAM signals. In topic (3), minimum required DAC speed is also investigated for generating Nyquist-WDM signals. The electrical filter shape and spectral pre-emphasis have a big impact on the system performances. By using a 4th-order super-Gaussian electrical filter and spectral pre-emphasis, 1.2-samples/symbol sampling speed is enough for DACs to generate PDM-16QAM Nyquist-WDM signals at 1.05-symbol-rate channel spacing, with OSNR penalties less than 0.4 dB for 28-Gbaud PDM-16QAM. In topic (3), a 2-channel interleaved DAC is also investigated to generated Nyquist-WDM signals, with zero-order holding at the speed of the symbol rate. The required spectral pre-emphasis, bandwidth and resolution are investigated. With proper spectral pre-emphasis, both the required electrical filter bandwidth and DAC resolution are similar to those with a conventional DAC used. Details are described in the context for the required parameters. A polyphase FIR filter structure is demonstrated in the transmitter to increase the processing parallelism and to reduce the calculation by half.

Topic (4) investigates the receiver-side matched filter design for RRC spectrally shaped Nyquist-WDM Systems. The required tap number of the receiver-side adaptive equalizer can be greatly reduced by adding matched filtering in the static chromatic dispersion equalizer. The matched filter makes a Nyquist-WDM system become more sensitive to the frequency offset between the lasers at the transmitter and receiver. Frequency-offset monitoring and compensation can reduce the frequency offset to a tolerable range. With a roll-off factor of 0.1, the OSNR penalty at BER 10^{-3} can be less than 0.2 dB from 1 GHz frequency offset.

Chapter 4 - An optical analogue of the well-established radio-frequency Yagi-Uda antenna, stands out by its effective unidirectional light emission and enhancement. Similar to radio wave and microwave antennas, their purpose is to convert the energy of free propagating radiation to localized energy, and vice versa. Optical antennas exploit the unique properties of metal nanostructures, which behave as strongly coupled plasmas at optical frequencies. Nobel metallic nanoparticles like Gold or silver with their plasmon resonances in the optical wavelength region can be used in a similar manner. Near-field coupling of quantum emitters to these particle plasmon resonances is used to redirect the light emission. Pursuing a brief introduction to the emerging field of optical nanoantennas, here the authors review the effect of changing the dipole, director and reflector's radii on the directivity of Yagi-Uda nanoantenna in near-field optical domain. The various aspects of the guiding properties of arrays of metamaterial or plasmonic particles, with a specific focus on Yagi-Uda nanoparticle models are investigated. In-depth analysis is performed on the transmission, reflection, and directive properties of nanoantenna.

In: Optical Communication Systems
Editor: Karla P. Boyd
ISBN: 978-1-63482-833-8

Chapter 1

MICRORING RESONATOR (MRR) OPTICAL SYSTEMS APPLIED TO ENHANCE THE SOLITON COMMUNICATIONS

I. S. Amiri[1,2]* and H. Ahmad[1]

[1] Photonics Research Centre, University of Malaya (UM), Kuala Lumpur, Malaysia

[2] Laser Center, Ibnu Sina ISIR, Universiti Teknologi Malaysia (UTM), Johor Bahru, Malaysia

ABSTRACT

A spectrum of light over a broad range can be generated, thus solitons can be utilized to generate chaotic filter features when spreading within nonlinear microring resonators (MRRs). We propose a systems of MRRs that uses localized soliton pulses to form the optical communication signals. This method uses nonlinear behaviours of light applicable for soliton transmission. Optical carrier generation is the basic building block for communication systems in the optical domain. One important aspect of the MRR system is that tuning of the system parameters allows for desired soliton carriers with specific key characteristics. The normalized output of the light field is defined by the ratio between the output and input fields. The results demonstrate optical soliton pulses localized within the MRRs system with 20,000 roundtrips.

* Email Address: amiri@um.edu.my.

A frequency soliton pulse can be formed and trapped within the panda ring resonator system with suitable ring parameters. The throughput output (E_{th}) shows localised soliton pulses with full width at half maximum (FWHM) and free spectral range (FSR) of 5 MHz and 2 GHz, respectively, where soliton pulses at frequencies of 50 and 52 GHz are generated. The drop output (E_d) shows pulses with FWHM and FSR of 10 MHz 2 GHz. The efficiency of this system can be evaluated by the output solitonic carrier signal generated by the MRRs.

Keywords: Microring Resonator (MRR), soliton, Kerr effect

1. Introduction

In this study, we propose a systems of MRRs that uses localized spatial and temporal soliton pulses to form the optical communication signals [1, 2]. The device parameters are considered with respect to the practical device parameters [3-5]. Soliton generation is very attractive especially when it uses quantum cryptography in an optical communication network, which is reported by Amiri *et al* [6-9]. A spectrum of light over a broad range can be generated by soliton pulses which are advised as a powerful laser pulses utilized to generate chaotic signals when spreading within microring resonators (MRRs) [10]. Therefore, the capacity of the transmission data can be secured and increased when the chaotic packet switching is utilized provided using this technique, [11-13]. Internet security becomes an important function in the modern internet service. However, the security technique known as quantum cryptography has been widely used and investigated in many applications, using ultra-short optical solitons [14-17]. Amiri *et al*, have proposed the new design of optical switching used in optical communication systems [18-22]. This method uses nonlinear behaviors of light in MRR which can be applicable for high-capacity transmission and switching [23-27]. Single-carrier modulation is a technology that modulates information onto only one carrier [28-30]. Modulation refers to the process of changing the carrier phase, frequency, amplitude, or their combination with a modulated signal that typically contains information to be transmitted [31, 32]. Orthogonal frequency division multiplexing (OFDM) is a combination of modulation and multiplexing of signals [33-36].

The main problem of this technology is satisfying the need for high bandwidth in fixed spectrum limits of one single-carrier [37-40]. High data rate in one carrier causes a high symbol rate. As the duration of one symbol or bit becomes smaller, the system becomes more susceptible to loss of information from impulse noise, signal reflections, and other impairments [41-45]. In addition, as the bandwidth used by a single-carrier system increases, the susceptibility to interference from other continuous signal sources becomes greater. This type of interference is commonly labelled as a carrier wave (CW) or frequency interference [46-49]. Bright-Gaussian soliton controls within a semiconductor add/drop multiplexer has numerous applications in optical communication [50, 51]. Optical soliton technique has become a powerful tool for many applications in science [52-60]. It can be used to trap and manipulate molecules/photons at mesoscopic scales with many applications in physical sciences [61-65]. The output is achieved when the high optical field is set up as an optical soliton [66-68]. For communication's application purposes, the optical soliton can be used to generate entangled photon within the proposed network system [69-73].

The MRR system provides a viable means to generate this building block that represents the generation of the necessary multi-carriers [74, 75]. Optical carrier generation thus constitutes the basic building block to implement OFDM transmission fully in the optical domain [76-79]. A soliton solution of the nonlinear wave equation is always stable over a long distance link, and this stability of soliton signals is even more remarkable than the possibility of balancing dispersion and non-linearity [80-83]. Transmission of all-optical OFDM can be implemented by generating multiple optical subcarriers, separating these subcarriers via optical devices, and finally modulating each subcarrier separately [84, 85]. One important aspect of the MRR system is that suitable tuning of the system parameters allows for desired soliton carriers to be obtained at the drop/through ports of the system [86-89].

2. Localized Pico/Femto Second Solitons within MRRs System

The suitable ring parameters are used, for instance, ring radii R_1=10 μm, R_2=5μm, and R_3=4μm. The schematic diagram of the proposed system is shown in Figure 1. A soliton pulse with peak power at 500 mW is input into the system. In order to make the system associate with the practical device, the selected parameters of the system are fixed to λ_0=1.55μm, n_0=3.34 (InGaAsP/InP) [90-92], A_{eff}=0.50, 0.25μm^2 and 0.12μm^2 for the different radii of microring resonators respectively [93-95], α=0.5dBmm^{-1}, γ=0.1 [96-98].

The coupling coefficient (κ) of the microring resonator ranged from 0.50 to 0.975 [99, 100].

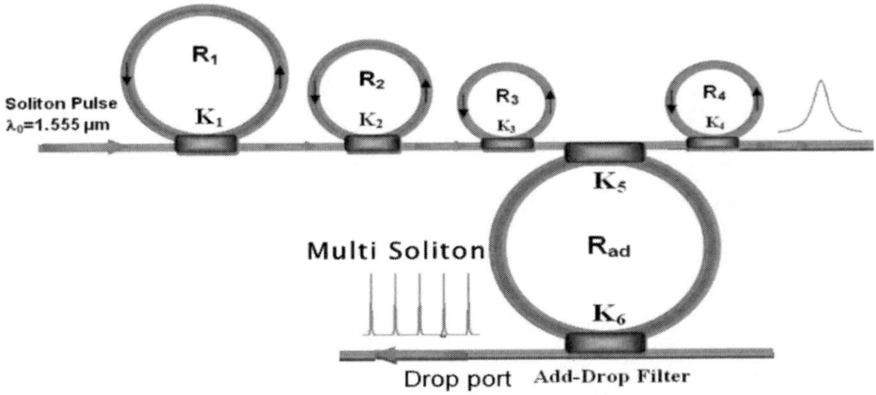

Figure 1. soliton pulse generation system, Rs, ring radii and κs, coupling coefficients.

Here, the bright soliton pulse is input into the system. The input optical field (E_{in}) of the bright soliton, dark soliton and Gaussian laser beam are given by:

$$E_{in} = A\sec h\left[\frac{T}{T_0}\right]\exp\left[\left(\frac{z}{2L_D}\right) - i\omega_0 t\right] \tag{1}$$

The soliton pulse is introduced into the proposed system. The input optical field (E_{in}) in the form of bright soliton is presented in Equation (1) [101-104]. A and z are the optical field amplitude and propagation distance, respectively [105, 106]. $L_D = T_0^2/|\beta_2|$ is the dispersion length of the soliton pulse [107]. The frequency shift of the soliton is ω_0. T is a soliton pulse propagation time in a frame moving at the group velocity , $T = t\text{-}\beta_1 \times z$, where β_1 and β_2 are the coefficients of the linear and second order terms of Taylor expansion of the propagation constant [108, 109].

This solution reports a pulse that maintains its temporal width invariance as it spreads, and thus is called a temporal soliton [110-112]. When a soliton peak intensity $\left(\left|\beta_2 / \Gamma T_0^2\right|\right)$ is established, then T_o is recognized [113, 114]. For the soliton pulse in the micro ring device, a balance should be accomplished between the dispersion length (L_D) and the nonlinear length ($L_{NL} = 1/\Gamma\varphi_{NL}$) [115], where $\Gamma = n_2 \times k_0$, is the length scale over which dispersive or nonlinear effects makes the beam gets wider or narrower [116, 117].

Since for a soliton pulse, there is a balance between dispersion and nonlinear lengths, therefore $L_D = L_{NL}$ [118]. Whenever light spreads within the nonlinear medium, the refractive index (n) of light within the medium is contributed by .

$$n = n_0 + n_2 I = n_0 + (\frac{n_2}{A_{eff}})P, \tag{2}$$

where n_0 and n_2 represent the linear and nonlinear refractive indices, respectively. I and P are the optical intensity and powers, respectively [119]. The effective mode core area of the device is introduced by A_{eff}. For the MRR and NRR, the effective mode core areas range 0.50 to 0.10 μm^2. When a soliton pulse is input and propagated within a microring resonator as shown in Figure 1, the resonant output can be performed, hence, the normalized output of the light field is defined by the ratio between the output and input fields presented by E_{out}(t) and E_{in}(t) in each roundtrip, and can be expressed as [120, 121]

$$\left|\frac{E_{out}(t)}{E_{in}(t)}\right|^2 = (1-\gamma)\left[1 - \frac{(1-(1-\gamma)x^2)\kappa}{(1-x\sqrt{1-\gamma}\sqrt{1-\kappa})^2 + 4x\sqrt{1-\gamma}\sqrt{1-\kappa}\sin^2(\frac{\phi}{2})}\right] \tag{3}$$

The equation (3) suggests that a microring resonator in the special case is very similar to a Fabry-Perot cavity, having an input and an output mirrors with a field reflectivity, (1-κ), and a fully reflecting mirror [122]. κ is the coupling coefficient, and $x = \exp(-\alpha L/2)$ symbolizes a roundtrip loss coefficient, $\Phi_0 = kLn_0$ and $\Phi_{NL} = kLn_2|E_{in}|^2$ are the linear and nonlinear phase shifts [123], $k = 2\pi/\lambda$ is the wave propagation number in a vacuum. Where L and α are the waveguide length and linear absorption coefficient, respectively [124, 125]. The nonlinear refractive index is n_2=2.2 x 10^{-17} m^2/W. The attenuation of the optical power within a MRR device is necessitated in order to maintain the constant output gain. In this case, the waveguide loss utilized is 0.5dBmm^{-1}.

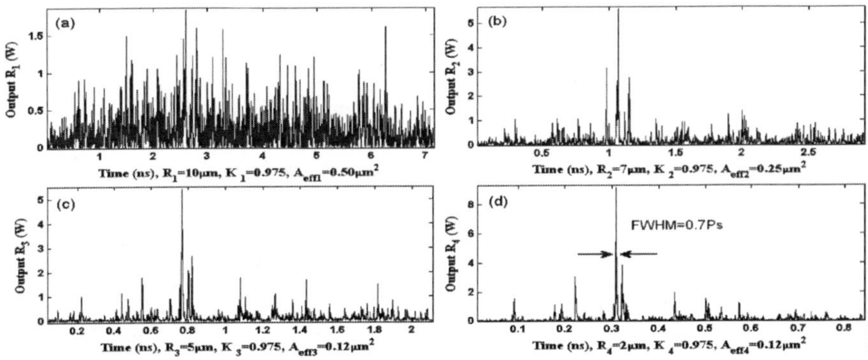

Figure 2. Results obtained when temporal soliton is localized within a microring device, where (a): Chaotic signals from R_1, (b): Chaotic signals from R_2, (c): temporal soliton, (d): temporal soliton with FWHM of 0.7 ps.

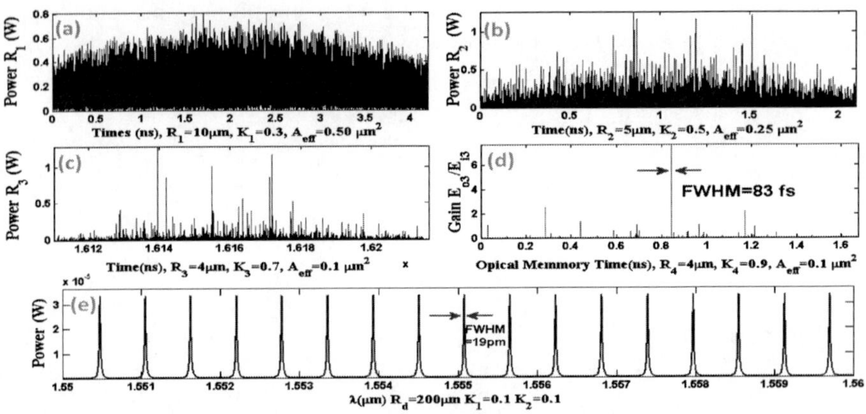

Figure 3. (a): Chaotic signals from R_1, (b): Chaotic signals from R_2, (c): filtering signals, (d): soliton with FWHM of 83 fs, (e): soliton with FWHM=19 pm.

Figure 3 shows the results while optical soliton pulses are localized within a MRRs and add/drop MRR system with 20,000 roundtrips, thus the optical pulse of 83 fs can be generated. Here, the ring radii are R_1= 10 μm, R_2= 5 μm, R_3= 4 μm, R_4=4 μm and R_{ad}= 200 μm with coupling coefficient of κ_1=0.3, κ_2=0.5, κ_3=0.7, κ_4=0.9, $\kappa_5 = \kappa_6 = 0.1$.

3. Frequency Domain Solitons for Communication Systems

Here, series of ring resonators are connected to a panda ring resonator system. The system of soliton frequency band generation is shown in Figure 4. The filtering process of the input soliton pulses is performed via the ring resonators, where frequency band ranges 40 to 60 GHz can be obtained via the output signals of the panda ring resonator system. The soliton pulse presented by equation (1) is introduced into the nonlinear system.

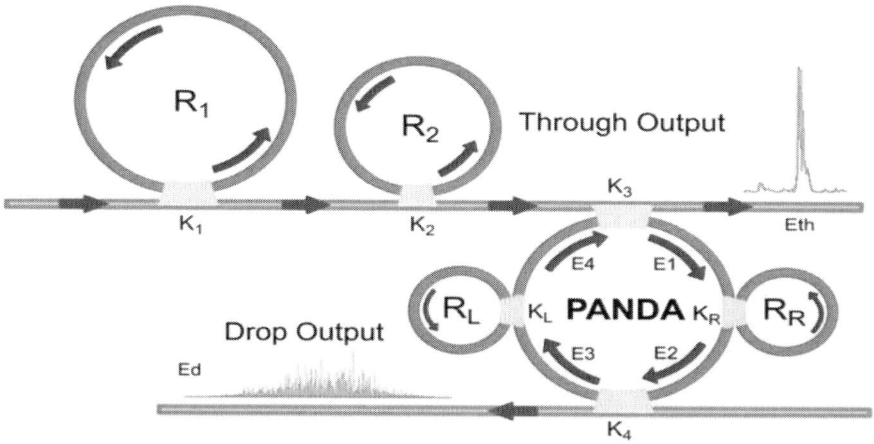

Figure 4. A panda MRR connected to a series of MRRs.

A frequency soliton pulse can be formed and trapped within the panda ring resonator system with suitable ring parameters. The centred ring of the panda ring resonator system has a radius of 100 μm and coupling coefficients of $\kappa_3 = 0.35$ and $\kappa_4 = 0.30$, where the right and left rings have radii of 18 μm and 8 μm, respectively. The effective core areas of the right and left rings are $A_{eff1} = A_{eff2} = 0.25\mu m^2$, where the coupling coefficients have been selected to $\kappa_R = 0.22$ and $\kappa_L = 0.10$. Filtering of the interior soliton signals can be performed when the pulses pass through the couplers, κ_3 and κ_4. The output signals from the throughput and drop ports of the system can be seen in Figure 5, where soliton range of 40 to 60 GHz are generated and used in many optical communication applications, such as wireless personal area networks (WPANs) and wireless local area networks (WLANs). The throughput output (E_{th}) shows localised ultra-short soliton pulses with an FWHM and free spectral range (FSR) of 5 MHz and 2 GHz, respectively, where soliton pulses at frequencies of 50 GHz and 52 GHz are generated. The drop output signals expressed by E_d are shown in Figure 5(b-c), where, pulses with FWHM of 10 MHz and FSR of 2 GHz are obtained.

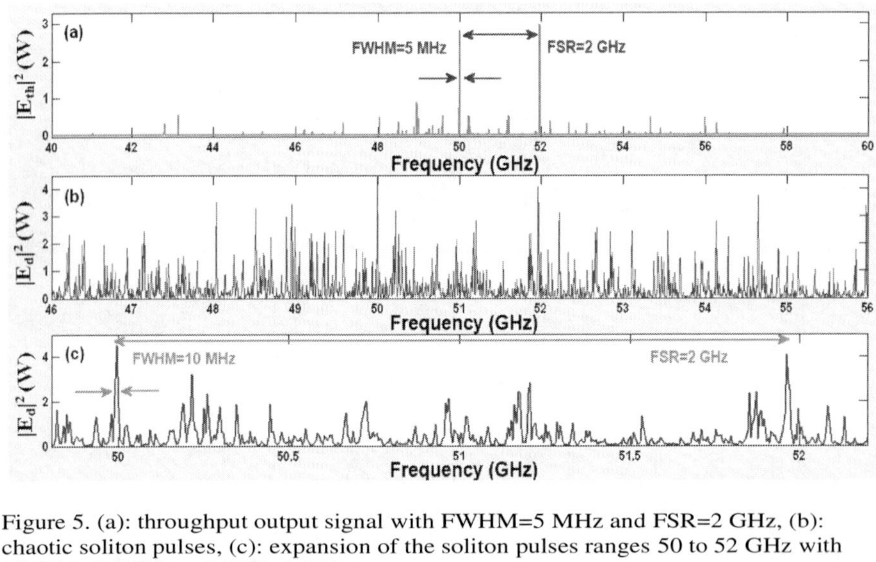

Figure 5. (a): throughput output signal with FWHM=5 MHz and FSR=2 GHz, (b): chaotic soliton pulses, (c): expansion of the soliton pulses ranges 50 to 52 GHz with FWHM=10 MHz and FSR=2 GHz.

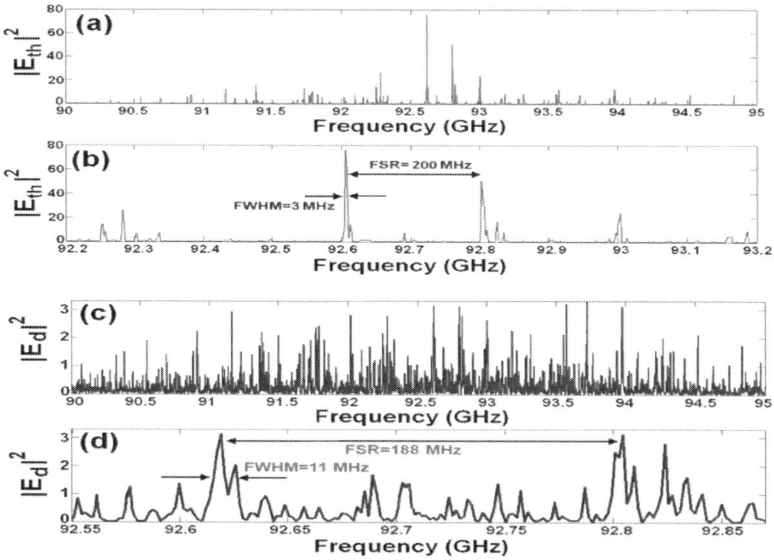

Figure 6. (a-b) throughput output signal (W/μm^2) with FWHM=3 MHz and FSR=200MHz, (c-d) drop port multi-solitons output signal (W/μm^2) with FWHM=11 MHz and FSR=188 MHz.

A bright soliton with a central frequency of 92.5 GHz and power of 2 W is introduced into the first MRR. The results of the chaotic signal generation are shown in Figure 6. The rings' radii and coupling coefficients are $R_1 = 6\mu m$, $\kappa_1 = 0.3$, $R_2 = 4\mu m$ and $\kappa_2 = 0.3$. The ring of the Panda ring resonator system has a radius of 120 μm and coupling coefficients of $\kappa_3 = 0.3$ and $\kappa_4 = 0.5$, where the right and left rings have the same radii of $R_R = 3\mu m$ and $R_L = 3\mu m$. The effective core areas of the rings are $A_{effR} = A_{effL} = 0.10\mu m^2$, where the coupling coefficients have been selected to $\kappa_R = 0.2$ and $\kappa_L = 0.1$. The wave-guided and fractional coupler intensity losses are $\alpha = 0.5$ dBmm^{-1}, and $\gamma = 0.1$ respectively. The through put output shows localized narrow bandwidth soliton pulses with FWHM and FSR of 3 MHz and 200 MHz, respectively, where soliton pulses at frequencies of 92.6, 92.8 and 93 GHz are generated. The drop port output signals are shown in Figure 6(c-d), where multi-soliton pulses with FWHM of 11 MHz and FSR of 188 MHz could be generated. As it can be concluded from the results, the efficiency of this system can be evaluated by the output solitonic carrier signal generated by the MRRs. Long distance fiber communication link is implemented using this method and can be integrated to short distance indoor wireless link for ultra-high data rate transmission.

4. Soliton Multi-Carries Generation Used in Optical Communication Systems

Using the system shown by Figure 1, the optical soliton carrier can be generated. Figure 7 shows the results when soliton pulses are localized within the microring device and add/drop microring resonator system with 20,000 roundtrips, where the soliton pulses with FWHM=20 MHz could be generated as result of the fourth ring resonator shown in Figure 7(d). The output results from the drop port of the system can be seen in the Figure 7(e-f). Here, the ring radii are R_1=15 μm, R_2=9 μm, R_3=6 μm, R_4=6 μm and R_{ad}=100 μm having coupling coefficients of κ_1=0.98, κ_2=0.98, κ_3=0.96, κ_4=0.92, κ_5=0.1 and κ_6=0.1. The nonlinear refractive index is 2.4×10^{-17}. A_{eff1}=0.50 μm^2, A_{eff2}=0.25 μm^2 and A_{eff3-4}=0.10 μm^2.

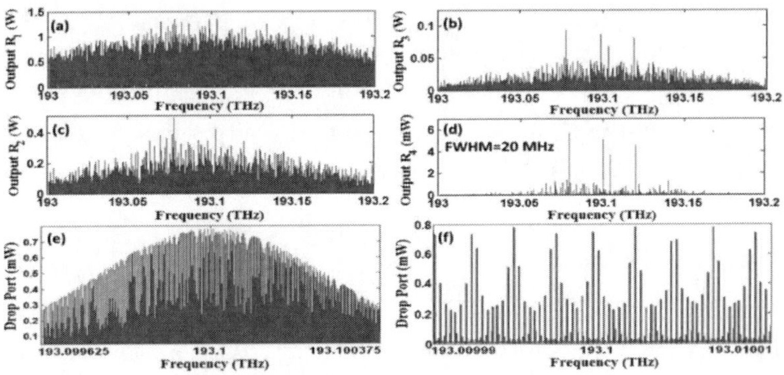

Figure 7. (a): output from first ring resonator, (b): output from third ring resonator, (c): output from second ring resonator, (d): output from fourth ring resonator, (e): multi-solitons output from the drop port, (f): expansion of the multi-carriers.

Thus, filtering of the input pulse within the system allowed for generation of single- and multi-soliton pulses.

CONCLUSION

The chaotic signal generation by means of a soliton pulse in the nonlinear microring resonators has been presented. We have proposed an interesting concept of the soliton pulse generation using microring resonators (MRRs). Soliton pulses could be achieved using the nonlinear condition established in the MRRs system. The balance based between the dispersion and nonlinear lengths of the soliton pulse presents the soliton behavior known as self-phase modulation, which introduces the optical output constant. It means that the light pulse can be localized coherently within the MRR system used in optical communication systems. Therefore, a large bandwidth of the soliton pulses can be generated and localized within a MRR system. Selected light pulse can be localized and used to perform the communication applications due to generate suitable bandwidth of the pulses. Localized soliton pulse are useful to generate optical communication signals applicable for network communications.

ACKNOWLEDGMENTS

Amiri and Ahmad would like to acknowledge the financial support from University Malaya/MOHE under grant number UM.C/625/1/HIR/MOHE /SCI/29 and RU002/2013.

REFERENCES

[1] A. Afroozeh, I.S. Amiri, K. Chaudhary, J. Ali & P. P. Yupapin, Analysis of Optical Ring Resonator, in *Advances in Laser and Optics Research*, ed New York: Nova Science, 2015.

[2] I. S. Amiri, J. Ali & P. P. Yupapin, (2012) "Enhancement of FSR and Finesse Using Add/Drop Filter and PANDA Ring Resonator Systems", *International Journal of Modern Physics B*, 26(04), 1250034.

[3] I. S. Amiri, S. E. Alavi & H. Ahmad, (2015) "Analytical Treatment of the Ring Resonator Passive Systems and Bandwidth Characterization

Using Directional Coupling Coefficients ", *Journal of Computational and Theoretical Nanoscience (CTN)*, 12(3),

[4] M. Bahadoran, I. S. Amiri, A. Afroozeh, J. Ali & P. P. Yupapin, (2011) "Analytical Vernier Effect for Silicon Panda Ring Resonator", presented at the *National Science Postgraduate Conference, NSPC* Universiti Teknologi Malaysia.

[5] I. S. Amiri, R. Ahsan, A. Shahidinejad, J. Ali & P. P. Yupapin, (2012) "Characterisation of bifurcation and chaos in silicon microring resonator", *IET Communications*, 6(16), 2671-2675.

[6] S. E. Alavi, I. S. Amiri, S. M. Idrus, ASM Supa'at, J. Ali & P. P. Yupapin, (2014) "All Optical OFDM Generation for IEEE802.11a Based on Soliton Carriers Using MicroRing Resonators ", *IEEE Photonics Journal*, 6(1),

[7] Iraj Sadegh Amiri & Harith Ahmad, Optical Soliton Communication Using Ultra-Short Pulses. USA: Springer, 2014.

[8] J. Ali, I. S. Amiri, A. Jalil, A. Kouhnavard, B. Mitatha & P. Yupapin, (2010) "Quantum internet via a quantum processor", presented at the *International Conference on Photonics (ICP 2010)*, Langkawi, Malaysia

[9] I. S. Amiri, A. Nikoukar, J. Ali & P. P. Yupapin, (2012) "Ultra-Short of Pico and Femtosecond Soliton Laser Pulse Using Microring Resonator for Cancer Cells Treatment", *Quantum Matter*, 1(2), 159-165.

[10] Amiri, H Ahmad & MZ Zulkifli, (2014) "Integrated ring resonator system analysis to Optimize the soliton transmission", *International Research Journal of Nanoscience and Nanotechnology*, 1(1), 002-007.

[11] I. S. Amiri, S. E. Alavi & J. Ali, (2013) "High Capacity Soliton Transmission for Indoor and Outdoor Communications Using Integrated Ring Resonators", *International Journal of Communication Systems*, 28(1), 147–160.

[12] IS Amiri & A Afroozeh, Integrated Ring Resonator Systems, in *Ring Resonator Systems to Perform Optical Communication Enhancement Using Soliton*, ed USA: Springer, 2014.

[13] D. Gifany, I. S. Amiri, M. Ranjbar & J. Ali, (2013) "LOGIC CODES GENERATION AND TRANSMISSION USING AN ENCODING-DECODING SYSTEM", *International Journal of Advances in Engineering & Technology (IJAET)*, 5(2), 37-45

[14] A. Afroozeh, I. S. Amiri, M. A. Jalil, M. Kouhnavard, J. Ali & P. P. Yupapin, (2011) "Multi Soliton Generation for Enhance Optical Communication", *Applied Mechanics and Materials*, 83 136-140.

[15] I. S. Amiri, (2011), "Optical Soliton Trapping for Quantum Key Generation", in *The International Conference for Nano materials Synthesis and Characterization*, Malaysia.

[16] IS Amiri & A Afroozeh, Soliton Generation Based Optical Communication, in *Ring Resonator Systems to Perform Optical Communication Enhancement Using Soliton*, ed USA: Springer, 2014.

[17] I. S. Amiri, S. E. Alavi, S. M. Idrus, A. S. M. Supa'at, J. Ali & P. P. Yupapin, (2014) "W-Band OFDM Transmission for Radio-over-Fiber link Using Solitonic Millimeter Wave Generated by MRR", *IEEE Journal of Quantum Electronics*, 50(8), 622 - 628.

[18] I. S. Amiri, A. Nikoukar & J. Ali, (2013) "GHz Frequency Band Soliton Generation Using Integrated Ring Resonator for WiMAX Optical Communication", *Optical and Quantum Electronics*, 46(9), 1165-1177.

[19] I. S. Amiri, S. E. Alavi, H. Ahmad, A.S.M. Supa'at & N. Fisal, (2014) "Numerical Computation of Solitonic Pulse Generation for Terabit/Sec Data Transmission", *Optical and Quantum Electronics*,

[20] I. S. Amiri & J. Ali, (2014) "Optical Quantum Generation and Transmission of 57-61 GHz Frequency Band Using an Optical Fiber Optics ", *Journal of Computational and Theoretical Nanoscience (CTN)*, 11(10), 2130-2135.

[21] I. S. Amiri, S. E. Alavi & S. M. Idrus, Theoretical Background of Microring Resonator Systems and Soliton Communication, in *Soliton Coding for Secured Optical Communication Link*, ed USA: Springer, 2015, pp. 17-39.

[22] I. S. Amiri, M. Ebrahimi, A. H. Yazdavar, S. Gorbani, S. E. Alavi, Sevia M. Idrus & J. Ali, (2014) "Transmission of data with orthogonal frequency division multiplexing technique for communication networks using GHz frequency band soliton carrier", *IET Communications*, 8(8), 1364 – 1373.

[23] I. S. Amiri, A. Shahidinejad, A. Nikoukar, M. Ranjbar, J. Ali & P. P. Yupapin, (2012) "Digital Binary Codes Transmission via TDMA Networks Communication System Using Dark and Bright Optical Soliton", *GSTF Journal on Computing (joc)*, 2(1), 12.

[24] A. A. Shojaei & I. S. Amiri, (2011) "DSA for Secured Optical Communication", presented at the *International Conference for Nanomaterials Synthesis and Characterization (INSC)*, Kuala Lumpur, Malaysia.

[25] I. S. Amiri, D. Gifany & J. Ali, (2013) "Long Distance Communication Using Localized Optical Soliton Via Entangled Photon", *IOSR Journal of Applied Physics (IOSR-JAP)*, 3(1), 32-39.

[26] I. S. Amiri, P. Naraei & J. Ali, (2014) "Review and Theory of Optical Soliton Generation Used to Improve the Security and High Capacity of MRR and NRR Passive Systems", *Journal of Computational and Theoretical Nanoscience (CTN)*, 11(9), 1875-1886.

[27] IS Amiri, MZ Zulkifli & H Ahmad, (2014) "Soliton comb generation using add-drop ring resonators", *International Research Journal of Telecommunications and Information Technology*, 1(1), 002-008.

[28] I. S. Amiri, A. Nikoukar, G. Vahedi, A. Shojaei, J. Ali & P. Yupapin, (2012) "Frequency-Wavelength Trapping by Integrated Ring Resonators For Secured Network and Communication Systems", *International Journal of Engineering Research and Technology (IJERT)*, 1(5),

[29] I. S. Amiri, A. Nikoukar, A. Shahidinejad, J. Ali & P. Yupapin, (2012), "Generation of discrete frequency and wavelength for secured computer networks system using integrated ring resonators", in *Computer and Communication Engineering (ICCCE) Conference*, Malaysia, 775-778.

[30] Abdolkarim Afroozeh, Iraj Sadegh Amiri & Alireza Zeinalinezhad, Micro Ring Resonators and Applications. Saarbrücken, Germany: LAP LAMBERT Academic Publishing, 2014.

[31] C. Tanaram, C. Teeka, R. Jomtarak, P. P. Yupapin, M. A. Jalil, I. S. Amiri & J. Ali, (2011) "ASK-to-PSK generation based on nonlinear microring resonators coupled to one MZI arm", *Procedia Engineering*, 8 432-435.

[32] S. E. Alavi, I. S. Amiri, S. M. Idrus & A. S. M. Supa'at, (2014) "Generation and Wired/Wireless Transmission of IEEE802.16m Signal Using Solitons Generated By Microring Resonator", *Optical and Quantum Electronics*,

[33] IS Amiri, SE Alavi, N Fisal, ASM Supa'at & H Ahmad, (2014) "All-Optical Generation of Two IEEE802.11n Signals for 2×2 MIMO-RoF via MRR System", *IEEE Photonics Journal*, 6(6),

[34] SE Alavi, IS Amiri, H Ahmad, ASM Supa'at & N Fisal, (2014) "Generation and Transmission of 3× 3 W-Band MIMO-OFDM-RoF Signals Using Micro-Ring Resonators", *Applied Optics*, 53(34), 8049-8054.

[35] I. S. Amiri, S. E. Alavi, Sevia M. Idrus, A. Nikoukar & J. Ali, (2013) "IEEE 802.15.3c WPAN Standard Using Millimeter Optical Soliton

Pulse Generated By a Panda Ring Resonator", *IEEE Photonics Journal*, 5(5), 7901912.

[36] Iraj Sadegh Amiri, Sayed Esan Alavi, Sevia Mahdaliza Idrus & Mojgan Kouhnavard, MICRORING RESONATOR FOR SECURED OPTICAL COMMUNICATION. USA: Amazon, 2014.

[37] I. S. Amiri, S. Ghorbani, P. Naraei & H. Ahmad, (2015) "Chaotic Carrier Signal Generation and Quantum Transmission Along Fiber Optics Communication Using Integrated Ring Resonators", *Quantum Matter*, 4(2),

[38] I. S. Amiri & J. Ali, (2013) "Data Signal Processing Via a Manchester Coding-Decoding Method Using Chaotic Signals Generated by a PANDA Ring Resonator", *Chinese Optics Letters*, 11(4), 041901(4).

[39] I. S. Amiri, K. Raman, A. Afroozeh, M. A. Jalil, I. N. Nawi, J. Ali & P. P. Yupapin, (2011) "Generation of DSA for security application", *Procedia Engineering*, 8 360-365.

[40] I. S. Amiri, S. E. Alavi, M. Bahadoran, A. Afroozeh & H. Ahmad, (2015) "Nanometer Bandwidth Soliton Generation and Experimental Transmission within Nonlinear Fiber Optics Using an Add-Drop Filter System", *Journal of Computational and Theoretical Nanoscience (CTN)*, 12(2),

[41] I. S. Amiri, S. Soltanmohammadi, A. Shahidinejad & j. Ali, (2013) "Optical quantum transmitter with finesse of 30 at 800-nm central wavelength using microring resonators", *Optical and Quantum Electronics*, 45(10), 1095-1105.

[42] M. Kouhnavard, I. S. Amiri, M. Jalil, A. Afroozeh, J. Ali & P. P. Yupapin, (2010) "QKD via a quantum wavelength router using spatial soliton", *AIP Conference Proceedings*, 1347 210-216.

[43] I. S. Amiri & A. Nikoukar, (2010-2011) "Quantum Information Generation Using Optical Potential Well", presented at the *Network Technologies & Communications (NTC) Conference*, Singapore.

[44] I. S. Amiri, S. E. Alavi & S. M. Idrus, Results of Digital Soliton Pulse Generation and Transmission Using Microring Resonators, in *Soliton Coding for Secured Optical Communication Link*, ed USA: Springer, 2015, pp. 41-56.

[45] I. S. Amiri & A. Nikoukar, (2010-2011) "Secured Binary Codes Generation for Computer Network Communication", presented at the *Network Technologies & Communications (NTC) Conference*, Singapore.

[46] I. S. Amiri, H. Ahmad & P. Naraei, (2015) "Optical Transmission Characteristics of an Optical Add-Drop Interferometer System", *Quantum Matter*, 4(5),
[47] A. Afroozeh, I. S. Amiri, M. Bahadoran, J. Ali & P. P. Yupapin, (2012) "Simulation of Soliton Amplification in Micro Ring Resonator for Optical Communication", *Jurnal Teknologi (Sciences and Engineering)*, 55 271-277.
[48] Iraj Sadegh Amiri, Ali Nikoukar & Sayed Ehsan Alavi, Soliton and Radio over Fiber (RoF) Applications. Saarbrücken, Germany: LAP LAMBERT Academic Publishing, 2014.
[49] I. S. Amiri, S. E. Alavi & S. M. Idrus', (2014) "Solitonic Pulse Generation and Characterization by Integrated Ring Resonators", presented at the *5th International Conference on Photonics 2014 (ICP2014)*, Kuala Lumpur.
[50] IS Amiri & A Afroozeh, Introduction of Soliton Generation, in *Ring Resonator Systems to Perform Optical Communication Enhancement Using Soliton*, ed USA: Springer, 2014.
[51] A. Nikoukar, I. S. Amiri, A. Shahidinejad, A. Shojaei, J. Ali & P. Yupapin, (2012), "MRR quantum dense coding for optical wireless communication system using decimal convertor", in *Computer and Communication Engineering (ICCCE) Conference*, Malaysia, 770-774.
[52] I. S. Amiri & J. Ali, (2014) "Deform of Biological Human Tissue Using Inserted Force Applied by Optical Tweezers Generated By PANDA Ring Resonator", *Quantum Matter*, 3(1), 24-28.
[53] S. E. Alavi, I. S. Amiri, M. R. K. Soltanian, A.S.M. Supa'at, N. Fisal & H. Ahmad, (2015) "Generation of Femtosecond Soliton Tweezers Using a Half-Panda System for Modeling the Trapping of a Human Red Blood Cell", *Journal of Computational and Theoretical Nanoscience (CTN)*, 12(1),
[54] I. S. Amiri & J. Ali, (2012) "Generation of Nano Optical Tweezers Using an Add/drop Interferometer System", presented at the *2nd Postgraduate Student Conference (PGSC)*, Singapore.
[55] A. Nikoukar, I. S. Amiri & J. Ali, (2013) "Generation of Nanometer Optical Tweezers Used for Optical Communication Networks", *International Journal of Innovative Research in Computer and Communication Engineering*, 1(1), 77-85.
[56] I. Sadegh Amiri, M. Nikmaram, A. Shahidinejad & J. Ali, (2013) "Generation of potential wells used for quantum codes transmission via

a TDMA network communication system", *Security and Communication Networks*, 6(11), 1301-1309.

[57] I. S. Amiri, A. Nikoukar, A. Shahidinejad, M. Ranjbar, J. Ali & P. P. Yupapin, (2012) "Generation of Quantum Photon Information Using Extremely Narrow Optical Tweezers for Computer Network Communication", *GSTF Journal on Computing (joc)*, 2(1), 140.

[58] I. S. Amiri & J. Ali, (2013) "Nano Particle Trapping By Ultra-short tweezer and wells Using MRR Interferometer System for Spectroscopy Application", *Nanoscience and Nanotechnology Letters*, 5(8), 850-856.

[59] I. S. Amiri & J. Ali, (2013) "Nano Optical Tweezers Generation Used for Heat Surgery of a Human Tissue Cancer Cells Using Add/Drop Interferometer System", *Quantum Matter*, 2(6), 489-493.

[60] I. S. Amiri & J. Ali, (2013) "Optical Buffer Application Used for Tissue Surgery Using Direct Interaction of Nano Optical Tweezers with Nano Cells", *Quantum Matter*, 2(6), 484-488.

[61] S. E. Alavi, I. S. Amiri, H. Ahmad, N. Fisal & ASM. Supa'at, (2015) "Optical Amplification of Tweezers and Bright Soliton Using an Interferometer Ring Resonator System", *Journal of Computational and Theoretical Nanoscience (CTN)*, 12(4),

[62] I. S. Amiri, B. Barati, P. Sanati, A. Hosseinnia, HR Mansouri Khosravi, S. Pourmehdi, A. Emami & J. Ali, (2014) "Optical Stretcher of Biological Cells Using Sub-Nanometer Optical Tweezers Generated by an Add/Drop Microring Resonator System", *Nanoscience and Nanotechnology Letters*, 6(2), 111-117.

[63] I. S. Amiri, A. Nikoukar, A Shahidinejad, T. Anwar & J. Ali, (2014) "Quantum Transmission of Optical Tweezers Via Fiber Optic Using Half-Panda System", *Life Science Journal*, 10(12s), 391-400.

[64] I. S. Amiri, A. Shahidinejad, A. Nikoukar, J. Ali & P. Yupapin, (2012) "A Study oF Dynamic Optical Tweezers Generation For Communication Networks", *International Journal of Advances in Engineering & Technology (IJAET)*, 4(2), 38-45

[65] N. Suwanpayak, S. Songmuang, M. A. Jalil, I. S. Amiri, I. Naim, J. Ali & P. P. Yupapin, (2010) "Tunable and storage potential wells using microring resonator system for bio-cell trapping and delivery", *AIP Conference Proceedings*, 1341 289-291.

[66] Iraj Sadegh Amiri & Abdolkarim Afroozeh, Ring Resonator Systems to Perform the Optical Communication Enhancement Using Soliton. USA: Springer, 2014.

[67] I. S. Amiri & J. Ali, (2013) "Single and Multi Optical Soliton Light Trapping and Switching Using Microring Resonator", *Quantum Matter*, 2(2), 116-121.

[68] A. A. Shojaei & I. S. Amiri, (2011) "Soliton for Radio wave generation", presented at the *International Conference for Nanomaterials Synthesis and Characterization (INSC)*, Kuala Lumpur, Malaysia.

[69] I. S. Amiri, G. Vahedi, A. Shojaei, A. Nikoukar, J. Ali & P. P. Yupapin, (2012) "Secured Transportation of Quantum Codes Using Integrated PANDA-Add/drop and TDMA Systems", *International Journal of Engineering Research and Technology (IJERT)*, 1(5),

[70] M. Kouhnavard, A. Afroozeh, M. A. Jalil, I. S. Amiri, J. Ali & P. P. Yupapin, (2010), "Soliton Signals and the Effect of Coupling Coefficient in MRR Systems", in *Faculty of Science Postgraduate Conference (FSPGC)*, Universiti Teknologi Malaysia.

[71] I. S. Amiri, D. Gifany & J. Ali, (2013) "Ultra-short Multi Soliton Generation for Application in Long Distance Communication", *Journal of Basic and Applied Scientific Research (JBASR)*, 3(3), 442-451.

[72] I. S. Amiri, A. Afroozeh, M. Bahadoran, J. Ali & P. P. Yupapin, (2011) "Up and Down Link of Soliton for Network Communication", presented at the *National Science Postgraduate Conference, NSPC*, Universiti Teknologi Malaysia.

[73] S. E. Alavi, I. S. Amiri, M. Khalily, A. S. M. Supa' at, N. Fisal, H. Ahmad & S. M. Idrus, (2014) "W-Band OFDM for Radio-over-Fibre Direct-Detection Link Enabled By Frequency Nonupling Optical Up-Conversion", *IEEE Photonics Journal* 6(6),

[74] I. S. Amiri & J. Ali, (2014) "Generating Highly Dark–Bright Solitons by Gaussian Beam Propagation in a PANDA Ring Resonator", *Journal of Computational and Theoretical Nanoscience (CTN)*, 11(4), 1092-1099.

[75] IS Amiri & A Afroozeh, Mathematics of Soliton Transmission in Optical Fiber, in *Ring Resonator Systems to Perform Optical Communication Enhancement Using Soliton*, ed USA: Springer, 2014.

[76] M. A. Jalil, I. S. Amiri, C. Teeka, J. Ali & P. P. Yupapin, (2011) "All-optical Logic XOR/XNOR Gate Operation using Microring and Nanoring Resonators", *Global Journal of Physics Express*, 1(1), 15-22.

[77] C. Teeka, S. Songmuang, R. Jomtarak, P. Yupapin, M. Jalil, I. S. Amiri & J. Ali, (2011) "ASK-to-PSK Generation based on Nonlinear Microring Resonators Coupled to One MZI Arm", *AIP Conference Proceedings*, 1341(1), 221-223.

[78] I. S. Amiri, S. E. Alavi & S. M. Idrus, Introduction of Fiber Waveguide and Soliton Signals Used to Enhance the Communication Security, in *Soliton Coding for Secured Optical Communication Link*, ed USA: Springer, 2015, pp. 1-16.

[79] I. S. Amiri & H. Ahmad, (2015) "Multiplex and De-multiplex of Generated Multi Optical Soliton By MRRs Using Fiber Optics Transmission Link", *Quantum Matter*, 4(4),

[80] A. Afroozeh, I. S. Amiri, A. Zeinalinezhad, S. E. Pourmand & H. Ahmad, (2015) "Comparison of Control Light using Kramers-Kronig Method by Three Waveguides", *Journal of Computational and Theoretical Nanoscience (CTN)*,

[81] I. S. Amiri, M. A. Jalil, A. Afroozeh, M. Kouhnavard , J. Ali & P. P. Yupapin, (2010), "Controlling Center Wavelength and Free Spectrum Range by MRR Radii", in *Faculty of Science Postgraduate Conference (FSPGC)*, Universiti Teknologi Malaysia.

[82] M. A. Jalil, I. S. Amiri, M. Kouhnavard, A. Afroozeh, J. Ali & P. P. Yupapin, (2010), "Finesse Improvements of Light Pulses within MRR System", in *Faculty of Science Postgraduate Conference (FSPGC)*, Universiti Teknologi Malaysia.

[83] I. S. Amiri, A. Afroozeh, J. Ali & P. P. Yupapin, (2012) "Generation Of Quantum Codes Using Up And Down Link Optical Solition", *Jurnal Teknologi (Sciences and Engineering)*, 55 97-106.

[84] S. E. Alavi, I.S.Amiri, A. S. M. Supa'at & S. M. Idrus, (2015) "Indoor Data Transmission Over Ubiquitous Infrastructure of Powerline Cables and LED Lighting", *Journal of Computational and Theoretical Nanoscience (CTN)*, 12(4),

[85] A. Shahidinejad, A. Nikoukar, I. S. Amiri, M. Ranjbar, A. Shojaei, J. Ali & P. Yupapin, (2012), "Network system engineering by controlling the chaotic signals using silicon micro ring resonator", in *Computer and Communication Engineering (ICCCE) Conference*, Malaysia, 765-769.

[86] IS Amiri, SE Alavi, A Shahidinejad, A Nikoukar, T Anwar, ASM Supa'at, SM Idrus & N. K. Yen, (2014) "Characterization of Ultra-Short Soliton Generation Using MRRs", presented at the *The 2014 Third ICT International Student Project Conference (ICT-ISPC2014)*, Thailand

[87] I. S. Amiri & J. Ali, (2013) "Controlling Nonlinear Behavior of a SMRR for Network System Engineering", *International Journal of Engineering Research and Technology (IJERT)*, 2(2),

[88] A. Afroozeh, I. S. Amiri, J. Ali & P. P. Yupapin, (2012) "Determination Of Fwhm For Solition Trapping", *Jurnal Teknologi (Sciences and Engineering)*, 55 77-83.

[89] IS Amiri, H. Ahmad & Hamza M. R. Al-Khafaji, (2015) "Full width at half maximum (FWHM) analysis of solitonic pulse applicable in optical network communication", *American Journal of Networks and Communications*, 4(1),

[90] I. S. Amiri, M. Nikmaram, A. Shahidinejad & J. Ali, (2012) "Cryptography Scheme of an Optical Switching System Using Pico/Femto Second Soliton Pulse", *International Journal of Advances in Engineering & Technology (IJAET)*, 5(1), 176-184.

[91] A. Afroozeh, M. Kouhnavard, I. S. Amiri, M. A. Jalil, J. Ali & P. P. Yupapin, (2010), "Effect of Center Wavelength on MRR Performance", in *Faculty of Science Postgraduate Conference (FSPGC)*, Universiti Teknologi Malaysia.

[92] Iraj Sadegh Amiri, Falah Jabar Rahim, Ari Sabir Arif, Sogand Ghorbani, Parisa Naraei, David Forsyth & Jalil Ali, (2014) "Single Soliton Bandwidth Generation and Manipulation by Microring Resonator", *Life Science Journal*, 10(12s), 904-910.

[93] I. S. Amiri, A. Afroozeh, I. N. Nawi, M. A. Jalil, A. Mohamad, J. Ali & P. P. Yupapin, (2011) "Dark Soliton Array for communication security", *Procedia Engineering*, 8 417-422.

[94] I. S. Amiri, D. Gifany & J. Ali, (2013) "Entangled Photon Encoding Using Trapping of Picoseconds Soliton pulse", *IOSR Journal of Applied Physics (IOSR-JAP)*, 3(1), 25-31.

[95] S. E. Alavi, I. S. Amiri, S. M. Idrus & J. Ali, (2013) "Optical Wired/Wireless Communication Using Soliton Optical Tweezers", *Life Science Journal*, 10(12s), 179-187.

[96] I. S. Amiri, G. Vahedi, A. Nikoukar, A. Shojaei, J. Ali & P. Yupapin, (2012) "Decimal Convertor Application for Optical Wireless Communication by Generating of Dark and Bright Signals of soliton", *International Journal of Engineering Research and Technology (IJERT)*, 1(5),

[97] I. S. Amiri, A. Afroozeh, M. Bahadoran, J. Ali & P. P. Yupapin, (2012) "Molecular Transporter System for Qubits Generation", *Jurnal Teknologi (Sciences and Engineering)*, 55 155-165.

[98] A. Afroozeh, I. S. Amiri, M. Kouhnavard, M. Jalil, J. Ali & P. Yupapin, (2010) "Optical dark and bright soliton generation and amplification", *AIP Conference Proceedings*, 1341 259-263.

[99] A. Afroozeh, A. Zeinalinezhad, SE. Pourmand & IS. Amiri, (2014) "Determination of Suitable Material to Control of Light", *International Journal of Biology, Pharmacy and Allied Sciences (IJBPAS)*, 3(11), 2410-2421.

[100] I. S. Amiri & J. Ali, (2014) "Picosecond Soliton pulse Generation Using a PANDA System for Solar Cells Fabrication", *Journal of Computational and Theoretical Nanoscience (CTN)*, 11(3), 693-701.

[101] A. Afroozeh, M. Bahadoran, I. S. Amiri, A. R. Samavati, J. Ali & P. P. Yupapin, (2011) "Fast Light Generation Using Microring Resonators for Optical Communication", presented at the *National Science Postgraduate Conference, NSPC*, Universiti Teknologi Malaysia.

[102] P. P. Yupapin, M. A. Jalil, I. S. Amiri, I. Naim & J. Ali, (2010) "New Communication Bands Generated by Using a Soliton Pulse within a Resonator System", *Circuits and Systems*, 1(2), 71-75.

[103] I. S. Amiri, A. Nikoukar, A. Shahidinejad & Toni Anwar, (2014) "The Proposal of High Capacity GHz Soliton Carrier Signals Applied for Wireless Commutation", *Reviews in Theoretical Science*, 2(4), 320-333.

[104] A. Afroozeh, A.Zeinalinezhad, I. S. Amiri & S. E. Pourmand, (2014) "Stop Light Generation using Nano Ring Resonators for ROM", *Journal of Computational and Theoretical Nanoscience (CTN)*, 12(3),

[105] Ali Shahidinejad, Iraj Sadegh Amiri & Toni Anwar, (2014) "Enhancement of Indoor Wavelength Division Multiplexing-Based Optical Wireless Communication Using Microring Resonator", *Reviews in Theoretical Science*, 2(3), 201-210.

[106] I. S. Amiri, A. Nikoukar & J. Ali, (2013) "Nonlinear Chaotic Signals Generation and Transmission Within an Optical Fiber Communication Link", *IOSR Journal of Applied Physics (IOSR-JAP)*, 3(1), 52-57.

[107] A. Afroozeh, M. Bahadoran, I. S. Amiri, A. R. Samavati, J. Ali & P. P. Yupapin, (2012) "Fast Light Generation Using GaAlAs/GaAs Waveguide", *Jurnal Teknologi (Sciences and Engineering)*, 57 17-23.

[108] M. Imran, R. A. Rahman & I. S. Amiri, (2010), "Fabrication of Diffractive Optical Element using Direct Writing CO2 Laser Irradiation", in *Faculty of Science Postgraduate Conference (FSPGC)*, Universiti Teknologi Malaysia.

[109] I. S. Amiri, A. Nikoukar & J. Ali, (2013) "New System of chaotic signal generation based on coupling coefficients applied to an Add/Drop System", *International Journal of Advances in Engineering & Technology (IJAET)*, 6(1), 78-87.

[110] I. S. Amiri & J. Ali, (2014) "Femtosecond Optical Quantum Memory generation Using Optical Bright Soliton", *Journal of Computational and Theoretical Nanoscience (CTN)*, 11(6), 1480-1485.

[111] I. S. Amiri, A. Afroozeh & M. Bahadoran, (2011) "Simulation and Analysis of Multisoliton Generation Using a PANDA Ring Resonator System", *Chinese Physics Letters*, 28(10), 104205.

[112] I. S. Amiri & A. Afroozeh, Spatial and Temporal Soliton Pulse Generation By Transmission of Chaotic Signals Using Fiber Optic Link in *Advances in Laser and Optics Research.* vol. 11, ed New York: Nova Science Publisher, 2015.

[113] I. S. Amiri, H. Ahmad & A. Shahidinejad, (2015) "Generating of 57-61 GHz Frequency Band Using a Panda Ring Resonator", *Quantum Matter*, 4(4),

[114] IS Amiri, SE Alavi, MRK Soltanian & H Ahmad, (2015) "Tunable Channel Spacing of Soliton Comb Generation Using Add-drop Microring Resonators (MRRs)", *Journal of Computational and Theoretical Nanoscience (CTN)*,

[115] A. Afrozeh, A. Zeinalinezhad, S. E. Pourmand & I. S. Amiri, (2014) "Attosecond Pulse Generation Using Nano Ring Waveguides", *INTERNATIONAL JOURNAL OF CURRENT LIFE SCIENCES*, 4(9), 7573-7575.

[116] S. E. Alavi, I. S. Amiri, S. M. Idrus, ASM. Supa'at & J. Ali, (2013) "Chaotic Signal Generation and Trapping Using an Optical Transmission Link", *Life Science Journal*, 10(9s), 186-192.

[117] Iraj Sadegh Amiri, Sayed Ehsan Alavi & Sevia Mahdaliza Idrus, Soliton Coding for Secured Optical Communication Link. USA: Springer, 2014.

[118] I. S. Amiri & J. Ali, (2014) "Simulation of the Single Ring Resonator Based on the Z-transform Method Theory", *Quantum Matter*, 3(6), 519-522.

[119] P. Sanati, A. Afroozeh, I. S. Amiri, J.Ali & Lee Suan Chua, (2014) "Femtosecond Pulse Generation using Microring Resonators for Eye Nano Surgery", *Nanoscience and Nanotechnology Letters*, 6(3), 221-226

[120] Iraj Sadegh Amiri, Sayed Ehsan Alavi, S. M. Idrus, Abdolkarim Afroozeh & Jalil Ali, Soliton Generation by Ring Resonator for Optical Communication Application. New York: Novascience Publishers, 2014.

[121] A. Shahidinejad, S. Soltanmohammadi, I. S. Amiri & T. Anwar, (2014) "Solitonic Pulse Generation for Inter-Satellite Optical Wireless Communication", *Quantum Matter*, 3(2), 150-154.

[122] I. S. Amiri, M. Ranjbar, A. Nikoukar, A. Shahidinejad, J. Ali & P. Yupapin, (2012), "Multi optical Soliton generated by PANDA ring resonator for secure network communication", in *Computer and Communication Engineering (ICCCE) Conference*, Malaysia, 760-764.

[123] A Nikoukar, IS Amiri, SE Alavi, A Shahidinejad, T Anwar, ASM Supa'at, SM Idrus & L. Y. Teng', (2014) "Theoretical and Simulation Analysis of The Add/Drop Filter Ring Resonator Based on the Z-transform Method Theory", presented at the *The 2014 Third ICT International Student Project Conference (ICT-ISPC2014)*, Thailand.

[124] I. S. Amiri, M. H. Khanmirzaei, M. Kouhnavard, P. P. Yupapin & J. Ali, Quantum Entanglement using Multi Dark Soliton Correlation for Multivariable Quantum Router, in *Quantum Entanglement* A. M. Moran, Ed., ed New York: Nova Science Publisher, 2012, pp. 111-122.

[125] A. Afroozeh, I. S. Amiri, A. Samavati, J. Ali & P. Yupapin, (2012), "THz frequency generation using MRRs for THz imaging", in *International Conference on Enabling Science and Nanotechnology (EsciNano)*, Kuala Lumpur, Malaysia, 1-2.

In: Optical Communication Systems
Editor: Karla P. Boyd
ISBN: 978-1-63482-833-8

Chapter 2

APPLICATION OF MICRORING RESONATORS (MRRS) IN OPTICAL SOLITON COMMUNICATIONS

***I. S. Amiri*[*,1,2], M. R. K. Soltanian*[1] *and H. Ahmad*[1]**
[1]Photonics Research Centre, University of Malaya, Malaysia
[2] Laser Center, Ibnu Sina ISIR, Universiti Teknologi Malaysia (UTM), Skudai, Johor Bahru, Malaysia

ABSTRACT

A system of bright soliton array (BSA) for optical communication using the wired/wireless link is presented. Gaussian laser beam with relevant parameters is input into the ring resonator system. Here, bright soliton with Full Width at Half Maximum (FWHM) and Free Specrum Range (FSR) of 120 ps and 370 ps are generated respectively. BSA can be obtained by a series of microring resonators (MRRs) with the input optical laser pulse. When a soliton pulse is propagated within a nonlinear microring device, a balance should be achieved between the dispersion length (LD) and the nonlinear length (NL), thus, the MRR in the particular case is very similar to an Fabry-Perot cavity, which has an input and output mirror with a field reflectivity. Beside improvements in efficiency and beam quality these soliton sources provide short and ultra-short pulses, leading to improved process efficiencies and new fields of

---

[*] Email Address: amiri@um.edu.my

optical soliton communications. In this study, transmission of BSA via 50 km fiber optics and 25 meters wireless has been investigated.

Keywords: center wavelength, array generation, multiplexer, ring resonator

1. Introduction

Fiber optic sensors and micro structured fibers hold great promise for integration of multiple sensing channels [1-3]. MRRs can be used by new applications in a wide range of nano photonics integrated systems [4-6]. To generate a spectrum of light over a broad range, an optical soliton pulse is recommended as a powerful laser pulse that can be used to generate chaotic filter characteristics [7-9]. Using this technique, the capacity of the transmission data can be secured and increased when the chaotic packet switching is employed [10-12]. Nonlinear behavior of light inside a MRR system takes place when a strong pulse of light is inserted into the ring system [13, 14]. Theoretical studies of such as systems have same conceptions with ring cavities, and Fabry–Perot system [15, 16]. Another technique used the nonlinear behavior of light in microring resonator to generate security codes [17-19].

In communication application, soliton can be used for long distance optical communication links [20]. For system performance, the required minimum repeater in the link is the key advantage [21-23]. To date, several papers have investigated soliton behaviors [24, 25]. In this study, the system consist of three MRRs is used to generate array of bright soliton in order to fulfil the communication security. The fundamental mechanism of soliton formation, namely, the balanced interplay of linear GVD and nonlinearity-induced self-phase modulation (SPM), is a well-understood concept [26-28]. The soliton pulses are so stable that the shape and velocity are preserved while travelling along the medium [29-31]. Besides improvements in efficiency and beam quality, these generated multi soliton provide short and ultra-short bandwidth, leading to improved process efficiencies and new fields of laser applications [32-35].

The main advantage of using soliton pulses in optical communication systems is that the shape of the pulse remains almost unaltered over a long distance [36-38]. In order to improve the system, narrower soliton pulses are recommended, where the attenuation of such signals during transmission lessens when compared to the conventional peaks of micrometre laser pulses

[39-41]. Generated optical pulses can be used for optical communications, in which the capacity of the output signals can be improved through the generation of peaks with smaller full width at half maximums (FWHMs) [42, 43]. To transmit the soliton signals via long distance communications, ultra-short soliton pulses are required [44-46].

2. Theoretical Simulation

According to the nonlinear Schrödinger equation (NLSE) model, the formation of a bright soliton from a large-amplitude wave packet is possible in systems with an attractive (or self-focusing) nonlinearity and is prohibited in systems with a repulsive (or defocusing) nonlinearity [47-49]. The underlying physics is as follows. The attractive nonlinearity produces a pulse self-narrowing effect; at a certain power level the self-narrowing can balance the dispersion-induced pulse self-broadening and give rise to the formation of a bright envelope soliton [50-52]. In contrast, in systems with a repulsive nonlinearity the nonlinearity induces self-broadening of the wave packet, just as the dispersiondoes, and thereby disables the formation of a bright soliton. The bright soliton array can be obtained using Gaussian pulse input to a series of microring resonator (MRR) in single system. The stationary Gaussian pulse is introduced into the MRR system, as shown in Figure 1.

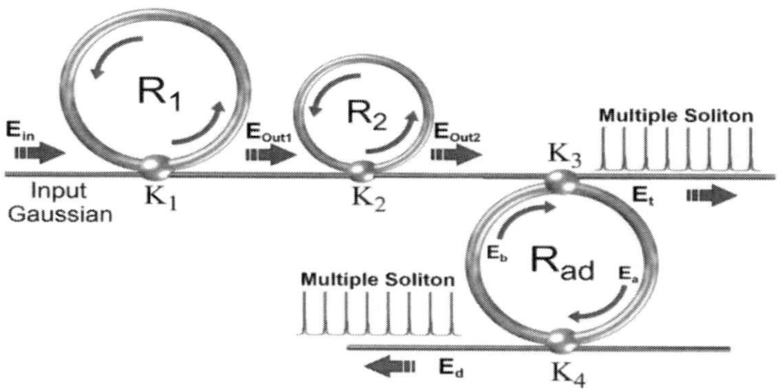

Figure 1. Schematic of bright soliton array generation. E_{in}, soliton inputs; R, ring radii; κ, coupling coefficients; MUX, optical multiplexer; R_d, add/drop radius; MRR, microring resonator.

Input optical fields of the Gaussian laser pulse is given by [53-55]

$$E_{in}(t) = E_0 \exp\left[(\frac{z}{2L_D}) - i\omega_0 t \right] \tag{1}$$

where A and P are the optical field amplitude and propagation distance, respectively. T is the soliton pulse propagation time in the frame moving at the group velocity, T = t-β_1*z, where $\beta_1$and β_2 are the coefficient of the linear and the second order terms of Taylor expansion of the propagation constant. $L_D = T_0^2 / |\beta_2|$ is the dispersion length of the soliton pulse [56-58]. T_o in the equation is a soliton pulse propagation time at initial input, where t is the soliton phase shift time, and the frequency shift of the soliton is ω_0. Equation (1) describes a pulse that keeps its temporal width invariance as it propagates and, thus, is called a temporal soliton [59-61]. When a soliton peak intensity $(|\beta_2 / \Gamma T_0^2|)$ is given, then T_0 is known. When a soliton pulse is propagated within a nonlinear microring device, a balance should be achieved between the dispersion length (L_D) and the nonlinear length ($L_{NL}=1/\Gamma\phi_{NL}$), where $\Gamma=n_2\times\kappa_0$ is the length scale over which dispersive or nonlinear effects make the beam wider or narrower [62-65]. For a soliton pulse, there is a balance between dispersion and nonlinear lengths, hence, $L_D = L_{NL}$. When light propagates within the nonlinear material (medium), the refractive index (n) of light within the medium is given by [66-68]

$$n = n_0 + n_2 I = n_0 + (\frac{n_2}{A_{eff}})P, \tag{2}$$

where n_0 and n_2 are the linear and nonlinear refractive indexes, respectively [69-71]. I and P are the optical intensity and optical power, respectively. The effective mode core area of the device is given by A_{eff} [72-75].

Thus, MRR in the particular case is very similar to a Fabry-Perot cavity, which has an input and output mirror with a field reflectivity, (1-κ), and a fully reflecting mirror [76-78]. κ and γ are the coupling coefficient and the fractional coupler intensity loss respectively, $x = \exp(-\alpha L/2)$ represents a round-trip loss coefficient, where the waveguide length and the linear absorption coefficient are given by L and α, respectively [79-81]. Here $\phi = \phi_0 + \varphi_{NL}$, where $\phi_0 = k_n L n_0$ and $\phi_{NL} = k_n L n_2 |E_{in}|^2$ are the linear and nonlinear phase shifts. Wave propagation number in a vacuum is given by $k_n = 2\pi / \lambda$, where λ is the input wavelength light field. For the add/drop system, the interior signals can be expressed by [82-84]:

$$E_a = E_{outR2} \times j\sqrt{\kappa_3} + E_b\sqrt{1-\kappa_3}\, e^{\frac{-\alpha}{2}\frac{L}{2} - jkn\frac{L}{2}} \tag{3}$$

$$E_b = E_a\sqrt{1-\kappa_4}\, e^{\frac{-\alpha}{2}\frac{L}{2} - jk_n\frac{L}{2}} \tag{4}$$

Thus [85-87],

$$E_a = \frac{E_{outR2} \times j\sqrt{\kappa_3}}{1 - \sqrt{1-\kappa_3}\sqrt{1-\kappa_4}\, e^{\frac{-\alpha}{2}L - jk_n L}} \tag{5}$$

$$E_b = \frac{E_{outR2} \times j\sqrt{\kappa_3}}{1 - \sqrt{1-\kappa_3}\sqrt{1-\kappa_4}\, e^{\frac{-\alpha}{2}L - jk_n L}} \cdot \sqrt{1-\kappa_4}\, e^{\frac{-\alpha}{2}\frac{L}{2} - jk_n\frac{L}{2}} \tag{6}$$

where κ_3 and κ_4 are the coupling coefficients, $L = 2\pi R$, and R is the radius of the add/drop filter device [88-90]. The throughput and drop port output electrical fields of the add/drop system can be expressed by [91-93]:

$$E_t = E_b e^{\frac{-\alpha}{2}\frac{L}{2} - jk_n\frac{L}{2}} j\sqrt{\kappa_3} + E_{outR2} \times \sqrt{1-\kappa_3} \tag{7}$$

$$E_d = E_a e^{\frac{-\alpha}{2}\frac{L}{2} - jk_n\frac{L}{2}} j\sqrt{\kappa_4} \tag{8}$$

Thus,

$$\frac{E_t}{E_{outR2}} = \frac{-\kappa_3\sqrt{1-\kappa_4}e^{\frac{-\alpha}{2}L-jk_nL} + \sqrt{1-\kappa_3} - (1-\kappa_3)\sqrt{1-\kappa_4}e^{\frac{-\alpha}{2}L-jk_nL}}{1-\sqrt{1-\kappa_3}\sqrt{1-\kappa_4}e^{\frac{-\alpha}{2}L-jk_nL}} = \frac{-\sqrt{1-\kappa_4}e^{\frac{-\alpha}{2}L-jk_nL} + \sqrt{1-\kappa_3}}{1-\sqrt{1-\kappa_3}\sqrt{1-\kappa_4}e^{\frac{-\alpha}{2}L-jk_nL}} \quad (9)$$

$$\frac{E_d}{E_{outR2}} = \frac{-\sqrt{\kappa_3.\kappa_4}e^{\frac{-\alpha}{2}\frac{L}{2}-jk_n\frac{L}{2}}}{1-\sqrt{1-\kappa_3}\sqrt{1-\kappa_4}e^{\frac{-\alpha}{2}L-jk_nL}} \quad (10)$$

The normalized outputs of the add/drop system are given by [94-97]:

$$|E_t|^2 = |E_{outR2}|^2 \times \frac{(1-\kappa_3) - 2\sqrt{1-\kappa_3}\cdot\sqrt{1-\kappa_4}\times e^{-\frac{\alpha}{2}L}\times\cos(k_nL) + (1-\kappa_4)\times e^{-\alpha L}}{1+(1-\kappa_3)(1-\kappa_4)\times e^{-\alpha L} - 2\sqrt{1-\kappa_3}\cdot\sqrt{1-\kappa_4}\times e^{-\frac{\alpha}{2}L}\times\cos(k_nL)} \quad (11)$$

$$|E_d|^2 = |E_{outR2}|^2 \times \frac{\kappa_3\kappa_4 \times e^{-\frac{\alpha}{2}L}}{1+(1-\kappa_3)(1-\kappa_4)\times e^{-\alpha L} - 2\sqrt{1-\kappa_3}\cdot\sqrt{1-\kappa_4}\times e^{-\frac{\alpha}{2}L}\times\cos(k_nL)} \quad (12)$$

Where E_t and E_d are the optical fields of the through and drop ports, respectively [98, 99]. The propagation constant is shown by $\beta=kn_{eff}$, where the effective refractive index of the waveguide is represented by n_{eff}. The circumference of the add/drop system is defined by $L=2\pi R$, where R is the radius of the ring [100-103]. By using particularly suitable parameters of the add/drop system, a cancellation of the chaotic signals can be carried out [104-107]. The MRR loss and the fractional coupler intensity loss are α=0.5dBmm^{-1} and γ=0.1, respectively [108-111].

3. Result and Discussion

The chaotic signals can be generated within the MRRs by using an input Gaussian laser pulse inserted into the waveguide. The proposed system is shown in Figure 1, where a Gaussian laser pulse with Full Width at Half Maximum (FWHM) of 20 ns, peak power 20 mW is inserted into the system. The appropriate ring parameters such as R_1=5 μm, R_2=3 μm, and R_{ad}=120 μm, λ_0=1.5 μm, n_0=3.34 (InGaAsP/InP), A_{eff1}=50 μm^2, A_{eff2}=25 μm^{2}andγ=0.1 are used. The coupling coefficient (κ) of the micro-ring resonatorsis κ_1=0.85, κ_2=0.80 where κ_3=0.06 and κ_4=0.06, where the nonlinear refractive index n_2=2.2×10^{-17} m^2/W. In this case, the waveguide loss is 0.5 dB mm^{-1}.

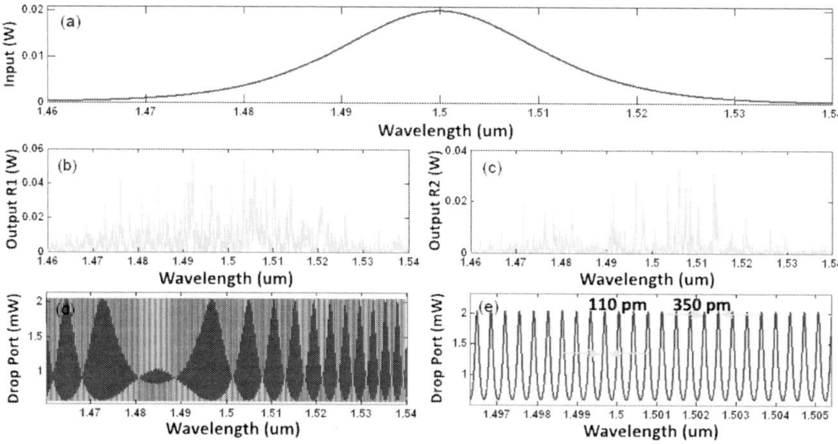

Figure 2. Spatial soliton pulse generation where (a): Input Gaussian laser pulse power, (b): Output result from first ring, (c): Output result from second ring, (d): Drop port output, (e): Expansion of the drop port output signals with FWHM=110 pm and FSR=350 pm.

Beside improvements in efficiency and beam quality these soliton sources provide short and ultra-short pulses, leading to improved process efficiencies and new fields of laser application [112-114]. The soliton pulses are so stable that its shape and velocity is preserved while travelling along the medium. The main advantage of using soliton pulses in optical communication systems is that the shape of the pulse remains almost unaltered over a long distance [115-117]. In order to improve the system, narrower soliton pulses are recommended where the attenuation of such signals during transmission become less comparatively to conventional peaks of micrometer laser pulses. The temporal shape of the optical soliton is shown in Figure 3.

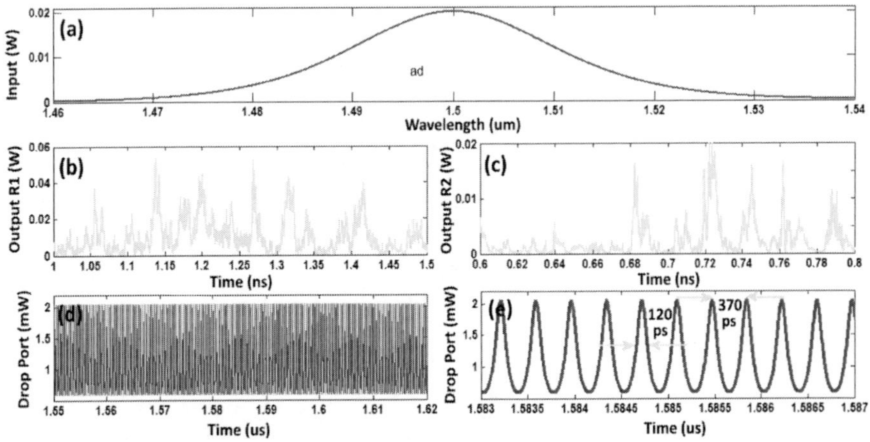

Figure 3. Temporal soliton pulse generation where (a): Input Gaussian laser pulse power, (b): Output result from first ring, (c): Output result from second ring, (d): Drop port output, (e): Expansion of the drop port output signals with FWHM=120 ps and FSR=370 ps.

In order to convert the multi bright soliton pulses, an optical circuit is presented shown in Figure 4.

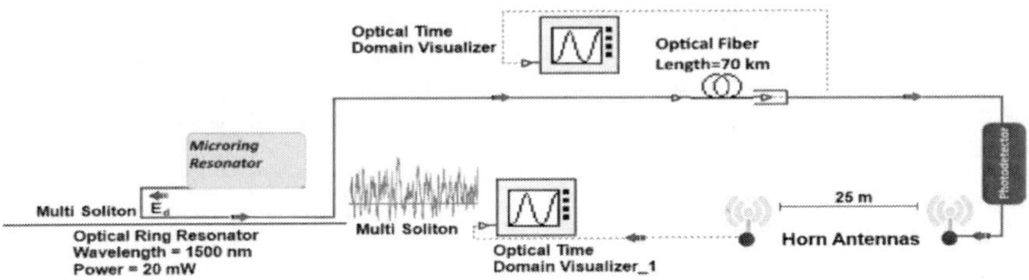

Figure 4. System setup.

The optical soliton pulses after generating by micro ring resonator (MRR) is then transmitted by optical fibre with the length of 70 km. The transmission of soliton array over a 70 km fibre is shown in Figure 5. The fiber optic has a length of 70 km, an attenuation of 0.2 dB/km, dispersion of 5 ps/nm/km, the differential group delay of 0.2 ps/km, the nonlinear refractive index of 2.6×10^{-20} m^2/W, effective area of 25 μm^2 and the nonlinear phase shift of 3 mrad. A

PIN photodetector with responsivity of 1 A/W and dark current of 10 nA is used to convert the optical pulses to array of radio frequency (RF). The centre frequency to carry the signals can be defined, or it can be calculated automatically by centering the filter at the optical channel with maximum power.

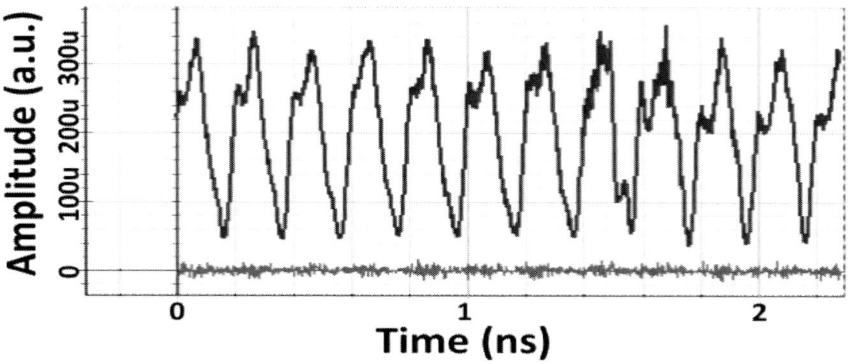

Figure 5. Transmission of soliton array over 70 km fibre optics.

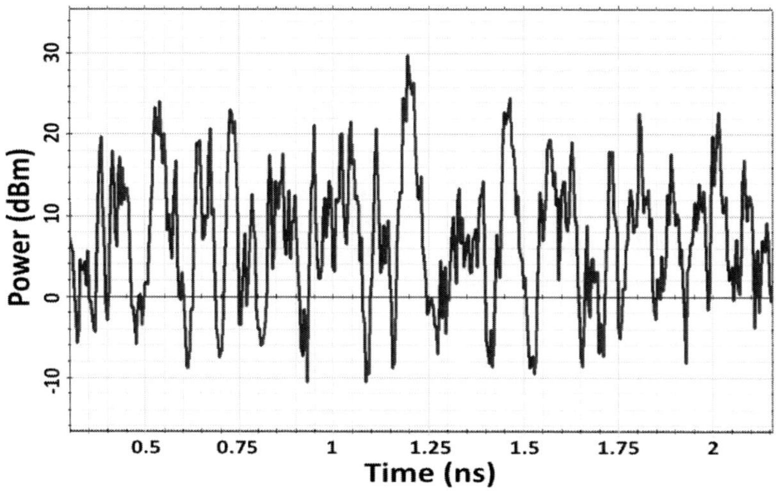

Figure 6.Transmission of bright soliton array over 25 m wireless link.

The sensitivity of a receiver can be defined by optimizing the receiver parameters. The transmitted soliton pulses received by the receiver over a 25 m wireless is shown in the Figure 6. Although exact optical soliton solutions have many potential applications in optical communications, laser sources or all-optical signal processing, and signal regeneration, they exist only under nondissipative conditions [118-120]. Unfortunately, in practical optical transmission systems, the intrinsic fiber loss makes unperturbed soliton transmission over long distances impossible even for low-loss media, as the integrity of the soliton pulse relies on the balance between the power-dependent self-phase modulation (SPM) in the transmission fiber and the constant anomalous chromatic dispersion of the fiber.

Conclusion

We have presented the use of ring resonators to generate soliton array in which the soliton pulses can be seen as results of the add/drop filter system. First, the Gaussian laser pulse launched into the series MRRs, through which multiple solitons with different center wavelengths, i.e., soliton arrays, can be obtained. Therefore, using the designed system, the soliton arrays pass through a 70 km optical fibre to connect a photodetector to convert optical soliton pulses to radio frequency (RF). The signals are the transmitted via 25 m wireless link and retrieve again.

Acknowledgments

The authors would like to acknowledge the financial support from University Malaya/MOHE under grant number UM.C/625/1/HIR/MOHE/SCI/29 and RU002/2013.

References

[1] A. Afrozeh, et al., Attosecond Pulse Generation Using Nano Ring Waveguides. *INTERNATIONAL JOURNAL OF CURRENT LIFE SCIENCES*, 2014. 4(9): p. 7573-7575.

[2] Iraj Sadegh Amiri, Abdolkarim Afroozeh, and Harith Ahmad, Integrated micro-ring photonics: Principles and Applications as Slow light devices, Soliton generation and optical transmission. 2015, United States: CRC Press.

[3] Amiri, I.S., S.E. Alavi, and S.M. Idrus, Theoretical Background of Microring Resonator Systems and Soliton Communication, in Soliton Coding for Secured Optical Communication Link. 2015, Springer: USA. p. 17-39.

[4] IS Amiri, MRK Soltanian, and H Ahmad, Application of Microring Resonators (MRRs) in Soliton Communication, in Optical Communication Systems: Fundamentals, Techniques and Applications. 2015, Novascience Publisher: New York.

[5] S. E. Alavi, et al., Cold Laser Therapy Modeling of Human Cell/Tissue by Soliton Tweezers. Optik, 2015.

[6] I. S. Amiri, et al., IEEE 802.15.3c WPAN Standard Using Millimeter Optical Soliton Pulse Generated By a Panda Ring Resonator. *IEEE Photonics Journal*, 2013. 5(5): p. 7901912.

[7] Amiri, I.S., et al. Multi optical Soliton generated by PANDA ring resonator for secure network communication. in Computer and Communication Engineering (ICCCE) Conference. 2012. Malaysia: IEEE Explore.

[8] Iraj Sadegh Amiri and Harith Ahmad, Optical Soliton Communication Using Ultra-Short Pulses. 2014, USA: Springer.

[9] I. S. Amiri, S. E. Alavi, and S.M. Idrus', Solitonic Pulse Generation and Characterization by Integrated Ring Resonators, in 5th International Conference on Photonics 2014 (ICP2014). 2014, *IEEE Explore: Kuala Lumpur*. p. 138 -140.

[10] I. S. Amiri, et al., Chaotic Carrier Signal Generation and Quantum Transmission Along Fiber Optics Communication Using Integrated Ring Resonators. *Quantum Matter*, 2015. 4(2): p. 151-155.

[11] S. E. Alavi, et al., Chaotic Signal Generation and Trapping Using an Optical Transmission Link. *Life Science Journal*, 2013. 10(9s): p. 186-192.

[12] Amiri, I.S., et al., Characterisation of bifurcation and chaos in silicon microring resonator. *IET Communications*, 2012. 6(16): p. 2671-2675.

[13] Teeka, C., et al., ASK-to-PSK Generation based on Nonlinear Microring Resonators Coupled to One MZI Arm. *AIP Conference Proceedings*, 2011. 1341(1): p. 221-223.

[14] Amiri, I.S. and J. Ali, Controlling Nonlinear Behavior of a SMRR for Network System Engineering. *International Journal of Engineering Research and Technology* (IJERT), 2013. 2(2).

[15] Jalil, M.A., et al., All-optical Logic XOR/XNOR Gate Operation using Microring and Nanoring Resonators. *Global Journal of Physics Express,* 2011. 1(1): p. 15-22.

[16] Afroozeh, A., et al., Fast Light Generation Using GaAlAs/GaAs Waveguide. *Jurnal Teknologi* (Sciences and Engineering), 2012. 57: p. 17-23.

[17] Amiri, I.S. and J. Ali, Data Signal Processing Via a Manchester Coding-Decoding Method Using Chaotic Signals Generated by a PANDA Ring Resonator. *Chinese Optics Letters*, 2013. 11(4): p. 041901(4).

[18] Amiri, I.S., et al., Generation Of Quantum Codes Using Up And Down Link Optical Solition. *Jurnal Teknologi* (Sciences and Engineering), 2012. 55: p. 97-106.

[19] Amiri, I.S., et al., Generation of Quantum Photon Information Using Extremely Narrow Optical Tweezers for Computer Network Communication. *GSTF Journal on Computing* (joc), 2012. 2(1): p. 140.

[20] Amiri, I.S., D. Gifany, and J. Ali, Long Distance Communication Using Localized Optical Soliton Via Entangled Photon. *IOSR Journal of Applied Physics* (IOSR-JAP), 2013. 3(1): p. 32-39.

[21] IS Amiri and H Ahmad, Mobility management in cellular radio networks, in Optical Communication Systems: *Fundamentals, Techniques and Applications*. 2015, Novascience Publisher: New York.

[22] A. Afroozeh, et al., Stop Light Generation using Nano Ring Resonators for ROM. *Journal of Computational and Theoretical Nanoscience* (CTN), 2014. 12(3).

[23] MRK Soltanian, et al., Yagi-Uda nanoantenna in the near-field optical domain, in Optical Communication Systems: Fundamentals, Techniques and Applications. 2015, Novascience Publisher: New York.

[24] A. Afroozeh, et al., Analysis of Optical Ring Resonator, in Advances in Laser and Optics Research. 2015, Nova Science: New York.

[25] A. Afroozeh, et al., Determination of Suitable Material to Control of Light. *International Journal of Biology, Pharmacy and Allied Sciences* (IJBPAS), 2014. 3(11): p. 2410-2421.

[26] I. S. Amiri, S. E. Alavi, and H. Ahmad, Analytical Treatment of the Ring Resonator Passive Systems and Bandwidth Characterization Using Directional Coupling Coefficients *Journal of Computational and Theoretical Nanoscience* (CTN), 2015. 12(3).

[27] Abdolkarim Afroozeh, et al., *Characterization and Controlling of Soliton Signals Generated by Semiconductor Microring Resonators*. 2015, USA: Springer.

[28] IS Amiri, et al., Tunable Channel Spacing of Soliton Comb Generation Using Add-drop Microring Resonators (MRRs). *Journal of Computational and Theoretical Nanoscience* (CTN), 2015.

[29] IS Amiri and A Afroozeh, Integrated Ring Resonator Systems, in Ring Resonator Systems to Perform Optical Communication Enhancement Using Soliton. 2014, Springer: USA.

[30] I. S. Amiri, H. Ahmad, and Hamza M. R. Al-Khafaji, A Review of Ultra-Short Soliton Pulse Generation Using InGaAsP/InP Microring Resonator (MRR) System. *American Journal of Networks and Communications*, , Special Issue: Recent Progresses in Optical Code-Division Multiple-Access (OCDMA) Technology, 2015. 4 (2-1): p. 6-17.

[31] Iraj Sadegh Amiri and Abdolkarim Afroozeh, Ring Resonator Systems to Perform the Optical Communication Enhancement Using Soliton, ed. L. DSilva. 2014, USA: Springer.

[32] Afroozeh, A., et al., Multi Soliton Generation for Enhance Optical Communication. *Applied Mechanics and Materials*, 2011. 83: p. 136-140.

[33] IS Amiri, et al., Multi Wavelength Mode-lock Soliton Generation Using Fiber Laser Loop Coupled to an Add-drop Ring Resonator. Optical and *Quantum Electronics*, 2015.

[34] I. S. Amiri and H. Ahmad, Multiplex and De-multiplex of Generated Multi Optical Soliton By MRRs Using Fiber Optics Transmission Link. *Quantum Matter*, 2015. 4(4).

[35] Amiri, I.S., A. Afroozeh, and M. Bahadoran, Simulation and Analysis of Multisoliton Generation Using a PANDA Ring Resonator System. Chinese Physics Letters, 2011. 28(10): p. 104205.

[36] IS Amiri and A Afroozeh, Introduction of Soliton Generation, in *Ring Resonator Systems to Perform Optical Communication Enhancement Using Soliton*. 2014, Springer: USA.

[37] Abdolkarim Afroozeh, Iraj Sadegh Amiri, and Alireza Zeinalinezhad, Micro Ring Resonators and Applications: New Technology of Optics, ed. A. Jian. 2014, Saarbrücken, Germany: LAP LAMBERT Academic Publishing.

[38] I. S. Amiri, P. Naraei, and J. Ali, Review and Theory of Optical Soliton Generation Used to Improve the Security and High Capacity of MRR

and NRR Passive Systems. *Journal of Computational and Theoretical Nanoscience* (CTN), 2014. 11(9): p. 1875-1886.

[39] Amiri, I.S., D. Gifany, and J. Ali, Ultra-short Multi Soliton Generation for Application in Long Distance Communication. *Journal of Basic and Applied Scientific Research* (JBASR), 2013. 3(3): p. 442-451.

[40] S. E. Alavi, et al., W-Band OFDM for Radio-over-Fibre Direct-Detection Link Enabled By Frequency Nonupling Optical Up-Conversion. *IEEE Photonics Journal* 2014. 6(6).

[41] I. S. Amiri, et al., W-Band OFDM Transmission for Radio-over-Fiber link Using Solitonic Millimeter Wave Generated by MRR. *IEEE Journal of Quantum Electronics, 2014*. 50(8): p. 622 - 628.

[42] IS Amiri, H. Ahmad, and Hamza M. R. Al-Khafaji, Full width at half maximum (FWHM) analysis of solitonic pulse applicable in optical network communication. *American Journal of Networks and Communications, , Special Issue: Recent Progresses in Optical Code-Division Multiple-Access (OCDMA) Technology*, 2015. 4(1).

[43] I. S. Amiri, et al., Transmission of data with orthogonal frequency division multiplexing technique for communication networks using GHz frequency band soliton carrier. *IET Communications*, 2014. 8(8): p. 1364 – 1373.

[44] IS Amiri, et al., All-Optical Generation of Two IEEE802.11n Signals for 2×2 MIMO-RoF via MRR System. *IEEE Photonics Journal*, 2014. 6(6).

[45] Amiri, H Ahmad, and MZ Zulkifli, Integrated ring resonator system analysis to Optimize the soliton transmission. International Research *Journal of Nanoscience and Nanotechnology*, 2014. 1(1): p. 002-007.

[46] I. S. Amiri, et al., Nanometer Bandwidth Soliton Generation and Experimental Transmission within Nonlinear Fiber Optics Using an Add-Drop Filter System. *Journal of Computational and Theoretical Nanoscience* (CTN), 2015. 12(2).

[47] Amiri, I.S., et al., Dark-Bright Solitons Conversion System for Secured and Long Distance Optical Communication. *IOSR Journal of Applied Physics* (IOSR-JAP), 2012. 2(1): p. 43-48.

[48] Afroozeh, A., et al., Optical dark and bright soliton generation and amplification. *AIP Conference Proceedings*, 2010. 1341: p. 259-263.

[49] Amiri, I.S. and J. Ali, Single and Multi Optical Soliton Light Trapping and Switching Using Microring Resonator. *Quantum Matter*, 2013. 2(2): p. 116-121.

[50] P. Sanati, et al., Femtosecond Pulse Generation using Microring Resonators for Eye Nano Surgery. *Nanoscience and Nanotechnology Letters,* 2014. 6(3): p. 221-226

[51] IS Amiri, MZ Zulkifli, and H Ahmad, Soliton comb generation using add-drop ring resonators. *International Research Journal of Telecommunications and Information Technology*, 2014. 1(1): p. 002-008.

[52] Amiri, I.S., et al., Ultra-Short of Pico and Femtosecond Soliton Laser Pulse Using Microring Resonator for Cancer Cells Treatment. *Quantum Matter*, 2012. 1(2): p. 159-165.

[53] I. S. Amiri, H. Ahmad, and A. Shahidinejad, Generating of 57-61 GHz Frequency Band Using a Panda Ring Resonator. *Quantum Matter*, 2015. 4(4).

[54] Amiri, I.S., S.E. Alavi, and S.M. Idrus, Introduction of Fiber Waveguide and Soliton Signals Used to Enhance the Communication Security, in *Soliton Coding for Secured Optical Communication Link*. 2015, Springer: USA. p. 1-16.

[55] IS Amiri and A Afroozeh, Mathematics of Soliton Transmission in Optical Fiber, in Ring Resonator Systems to Perform Optical Communication Enhancement Using Soliton. 2014, Springer: USA.

[56] S. E. Alavi, et al., Generation and Wired/Wireless Transmission of IEEE802.16m Signal Using Solitons Generated By Microring Resonator. *Optical and Quantum Electronics*, 2014.

[57] Yupapin, P.P., et al., New Communication Bands Generated by Using a Soliton Pulse within a Resonator System. *Circuits and Systems*, 2010. 1(2): p. 71-75.

[58] S. E. Alavi, et al., Optical Amplification of Tweezers and Bright Soliton Using an Interferometer Ring Resonator System. *Journal of Computational and Theoretical Nanoscience* (CTN), 2015. 12(4).

[59] Amiri, I.S., et al., Generation of DSA for security application. *Procedia Engineering,* 2011. 8: p. 360-365.

[60] Shahidinejad, A., et al. Network system engineering by controlling the chaotic signals using silicon micro ring resonator. in *Computer and Communication Engineering* (ICCCE) Conference. 2012. Malaysia: IEEE Explore.

[61] S. E. Alavi, et al., Optical Wired/Wireless Communication Using Soliton Optical Tweezers. *Life Science Journal*, 2013. 10(12s): p. 179-187.

[62] S. E. Alavi, et al., Generation of Femtosecond Soliton Tweezers Using a Half-Panda System for Modeling the Trapping of a Human Red Blood

Cell. *Journal of Computational and Theoretical Nanoscience* (CTN), 2015. 12(1): p. 10-18.

[63] S. E. Alavi, et al., Indoor Data Transmission Over Ubiquitous Infrastructure of Powerline Cables and LED Lighting. *Journal of Computational and Theoretical Nanoscience* (CTN), 2015. 12(4).

[64] I. S. Amiri, H. Ahmad, and P. Naraei, Optical Transmission Characteristics of an Optical Add-Drop Interferometer System. *Quantum Matter,* 2015. 4(5).

[65] Iraj Sadegh Amiri, Ali Nikoukar, and Sayed Ehsan Alavi, Soliton and Radio over Fiber (RoF) Applications, ed. A. Jian. 2014, Saarbrücken, Germany: LAP LAMBERT Academic Publishing.

[66] Nikoukar, A., I.S. Amiri, and J. Ali, Generation of Nanometer Optical Tweezers Used for Optical Communication Networks. *International Journal of Innovative Research in Computer and Communication Engineering*, 2013. 1(1): p. 77-85.

[67] Amiri, I.S., et al., Molecular Transporter System for Qubits Generation. *Jurnal Teknologi* (Sciences and Engineering), 2012. 55: p. 155-165.

[68] Amiri, I.S., A. Nikoukar, and J. Ali, Nonlinear Chaotic Signals Generation and Transmission Within an Optical Fiber Communication Link. *IOSR Journal of Applied Physics* (IOSR-JAP), 2013. 3(1): p. 52-57.

[69] Amiri, I.S. and J. Ali, Optical Buffer Application Used for Tissue Surgery Using Direct Interaction of Nano Optical Tweezers with Nano Cells. *Quantum Matter*, 2013. 2(6): p. 484-488.

[70] I. S. Amiri, S. E. Alavi, and H. Ahmad, Radio Frequency signal generation and wireless transmission using PANDA and Add/drop systems. *Journal of Computational and Theoretical Nanoscience* (CTN), 2015. 12(8).

[71] Iraj Sadegh Amiri, Sayed Ehsan Alavi, and Sevia Mahdaliza Idrus, Soliton Coding for Secured Optical Communication Link, ed. D. Loy. 2014, USA: Springer.

[72] Gifany, D., et al., LOGIC CODES GENERATION AND TRANSMISSION USING AN ENCODING-DECODING SYSTEM. *International Journal of Advances in Engineering & Technology* (IJAET), 2013. 5(2): p. 37-45

[73] Iraj Sadegh Amiri, et al., *MICRORING RESONATOR FOR SECURED OPTICAL COMMUNICATION*. 2014, USA: Amazon.

[74] Kouhnavard, M., et al., QKD via a quantum wavelength router using spatial soliton. *AIP Conference Proceedings*, 2010. 1347: p. 210-216.

[75] Amiri, I.S., et al., Quantum Entanglement using Multi Dark Soliton Correlation for Multivariable Quantum Router, in Quantum Entanglement A.M. Moran, Editor. 2012, Nova Science Publisher: New York. p. 111-122.

[76] I. S. Amiri and J. Ali, Femtosecond Optical Quantum Memory generation Using Optical Bright Soliton. *Journal of Computational and Theoretical Nanoscience* (CTN), 2014. 11(6): p. 1480-1485.

[77] I. S. Amiri and J. Ali, Nano Particle Trapping By Ultra-short tweezer and wells Using MRR Interferometer System for Spectroscopy Application. *Nanoscience and Nanotechnology Letters*, 2013. 5(8): p. 850-856.

[78] I. S. Amiri, et al., Quantum Transmission of Optical Tweezers Via Fiber Optic Using Half-Panda System. *Life Science Journal*, 2014. 10(12s): p. 391-400.

[79] Amiri, I.S., et al., Frequency-Wavelength Trapping by Integrated Ring Resonators For Secured Network and Communication Systems. *International Journal of Engineering Research and Technology* (IJERT), 2012. 1(5).

[80] I. S. Amiri and J. Ali, Optical Quantum Generation and Transmission of 57-61 GHz *Frequency Band Using an Optical Fiber Optics Journal of Computational and Theoretical Nanoscience* (CTN), 2014. 11(10): p. 2130-2135.

[81] Afroozeh, A., et al., Simulation of Soliton Amplification in Micro Ring Resonator for Optical Communication. *Jurnal Teknologi* (Sciences and Engineering), 2012. 55: p. 271-277.

[82] Amiri, I.S. and J. Ali, Deform of Biological Human Tissue Using Inserted Force Applied by Optical Tweezers Generated By PANDA Ring Resonator. *Quantum Matter,* 2014. 3(1): p. 24-28.

[83] Amiri, I.S. and J. Ali, Nano Optical Tweezers Generation Used for Heat Surgery of a Human Tissue Cancer Cells Using Add/Drop Interferometer System. *Quantum Matter*, 2013. 2(6): p. 489-493.

[84] I. S. Amiri, et al., Optical Stretcher of Biological Cells Using Sub-Nanometer Optical Tweezers Generated by an Add/Drop Microring Resonator System. *Nanoscience and Nanotechnology Letters*, 2014. 6(2): p. 111-117.

[85] SE Alavi, et al., Generation and Transmission of 3× 3 W-Band MIMO-OFDM-RoF Signals Using Micro-Ring Resonators. *Applied Optics,* 2014. 53(34): p. 8049-8054.

[86] I. S. Amiri, et al., Optical quantum transmitter with finesse of 30 at 800-nm central wavelength using microring resonators. *Optical and Quantum Electronics,* 2013. 45(10): p. 1095-1105.

[87] IS Amiri and A Afroozeh, Soliton Generation Based Optical Communication, in Ring Resonator Systems to Perform Optical Communication Enhancement Using Soliton. 2014, Springer: USA.

[88] Amiri, I.S., et al. Generation of discrete frequency and wavelength for secured computer networks system using integrated ring resonators. in *Computer and Communication Engineering* (ICCCE) Conference. 2012. Malaysia: IEEE Explore.

[89] Amiri, I.S., A. Nikoukar, and J. Ali, New System of chaotic signal generation based on coupling coefficients applied to an Add/Drop System. *International Journal of Advances in Engineering & Technology* (IJAET), 2013. 6(1): p. 78-87.

[90] Amiri, I.S., S.E. Alavi, and S.M. Idrus, Results of Digital Soliton Pulse Generation and Transmission Using Microring Resonators, in *Soliton Coding for Secured Optical Communication Link*. 2015, Springer: USA. p. 41-56.

[91] S. E. Alavi, et al., All Optical OFDM Generation for IEEE802.11a *Based on Soliton Carriers Using MicroRing Resonators IEEE Photonics Journal*, 2014. 6(1).

[92] I. S. Amiri and J. Ali, Generating Highly Dark–Bright Solitons by Gaussian Beam Propagation in a PANDA Ring Resonator. *Journal of Computational and Theoretical Nanoscience* (CTN), 2014. 11(4): p. 1092-1099.

[93] Amiri, I.S., et al., Secured Transportation of Quantum Codes Using Integrated PANDA-Add/drop and TDMA Systems. *International Journal of Engineering Research and Technology* (IJERT), 2012. 1(5).

[94] Rabus, D.G., Realization of optical filters using ring resonators with integrated semiconductor optical amplifiers in GaInAsP/InP. 2002, Universitätsbibliothek.

[95] Tanaram, C., et al., ASK-to-PSK generation based on nonlinear microring resonators coupled to one MZI arm. *Procedia Engineering,* 2011. 8: p. 432-435.

[96] Amiri, I.S., J. Ali, and P.P. Yupapin, Enhancement of FSR and Finesse Using Add/Drop Filter and PANDA Ring Resonator Systems. *International Journal of Modern Physics B,* 2012. 26(04): p. 1250034.

[97] I. S. Amiri, S. E. Alavi, and J. Ali, High Capacity Soliton Transmission for Indoor and Outdoor Communications Using Integrated Ring

Resonators. *International Journal of Communication Systems*, 2013. 28(1): p. 147–160.

[98] Amiri, I.S., D. Gifany, and J. Ali, Entangled Photon Encoding Using Trapping of Picoseconds Soliton pulse. *IOSR Journal of Applied Physics* (IOSR-JAP), 2013. 3(1): p. 25-31.

[99] Sadegh Amiri, I., et al., Generation of potential wells used for quantum codes transmission via a TDMA network communication system. *Security and Communication Networks*, 2013. 6(11): p. 1301-1309.

[100] Amiri, I.S. and J. Ali, Characterization of Optical Bistability In a Fiber Optic Ring Resonator. *Quantum Matter*, 2014. 3(1): p. 47-51.

[101] Amiri, I.S., et al., Dark Soliton Array for communication security. *Procedia Engineering*, 2011. 8: p. 417-422.

[102] Afroozeh, A., et al., Determination Of Fwhm For Solition Trapping. *Jurnal Teknologi (Sciences and Engineering),* 2012. 55: p. 77-83.

[103] I. S. Amiri, A. Nikoukar, and J. Ali, GHz Frequency Band Soliton Generation Using Integrated Ring Resonator for WiMAX Optical Communication. *Optical and Quantum Electronics*, 2013. 46(9): p. 1165-1177.

[104] A. Afroozeh, et al., Comparison of Control Light using Kramers-Kronig Method by Three Waveguides. *Journal of Computational and Theoretical Nanoscience (CTN),* 2015. 12(8).

[105] Amiri, I.S., et al., Cryptography Scheme of an Optical Switching System Using Pico/Femto Second Soliton Pulse. *International Journal of Advances in Engineering & Technology (IJAET),* 2012. 5(1): p. 176-184.

[106] Amiri, I.S., et al., Decimal Convertor Application for Optical Wireless Communication by Generating of Dark and Bright Signals of soliton. *International Journal of Engineering Research and Technology* (IJERT), 2012. 1(5).

[107] Iraj Sadegh Amiri, et al., Soliton Generation by Ring Resonator for Optical Communication Application, ed. J. Ramirez. 2014, New York: Novascience Publishers.

[108] Filho, A.F.G.F., et al., A performance study of a nonlinear all Fibre Michelson interferometer, add-drop multiplexer, based in Fibre Bragg grating mirrors. *Optical and quantum electronics*, 2008. 40(7): p. 525-534.

[109] Amiri, I.S., et al., Digital Binary Codes Transmission via TDMA Networks Communication System Using Dark and Bright Optical Soliton. *GSTF Journal on Computing (joc),* 2012. 2(1): p. 12.

[110] Nikoukar, A., et al. MRR quantum dense coding for optical wireless communication system using decimal convertor. in Computer and Communication Engineering (ICCCE) Conference. 2012. Malaysia: IEEE Explore.
[111] I. S. Amiri and A. Afroozeh, Spatial and Temporal Soliton Pulse Generation By Transmission of Chaotic Signals Using Fiber Optic Link in Advances in Laser and Optics Research. 2015, Nova Science Publisher: New York.
[112] Iraj Sadegh Amiri, et al., Single Soliton Bandwidth Generation and Manipulation by Microring Resonator. *Life Science Journal*, 2014. 10(12s): p. 904-910.
[113] A. Shahidinejad, et al., Solitonic Pulse Generation for Inter-Satellite Optical Wireless Communication. *Quantum Matter*, 2014. 3(2): p. 150-154.
[114] Suwanpayak, N., et al., Tunable and storage potential wells using microring resonator system for bio-cell trapping and delivery. *AIP Conference Proceedings*, 2010. 1341: p. 289-291.
[115] Ali Shahidinejad, Iraj Sadegh Amiri, and Toni Anwar, Enhancement of Indoor Wavelength Division Multiplexing-Based Optical Wireless Communication Using Microring Resonator. *Reviews in Theoretical Science*, 2014. 2(3): p. 201-210.
[116] I. S. Amiri, et al., Numerical Computation of Solitonic Pulse Generation for Terabit/Sec Data Transmission. *Optical and Quantum Electronics*, 2014.
[117] Amiri, I.S. and J. Ali, Picosecond Soliton pulse Generation Using a PANDA System for Solar Cells Fabrication. *Journal of Computational and Theoretical Nanoscience (CTN),* 2014. 11(3): p. 693-701.
[118] I. S. Amiri, et al., The Proposal of High Capacity GHz Soliton Carrier Signals Applied for Wireless Commutation. *Reviews in Theoretical Science,* 2014. 2(4): p. 320-333.
[119] I. S. Amiri and J. Ali, Simulation of the Single Ring Resonator Based on the Z-transform Method Theory. *Quantum Matter*, 2014. 3(6): p. 519-522.
[120] Amiri, I.S., et al., A Study oF Dynamic Optical Tweezers Generation For Communication Networks. *International Journal of Advances in Engineering & Technology (IJAET), 2012*. 4(2): p. 38-45

In: Optical Communication Systems
Editor: Karla P. Boyd
ISBN: 978-1-63482-833-8

Chapter 3

DIGITAL SIGNAL PROCESSING FOR NYQUIST-WDM OPTICAL FIBER TRANSMISSION SYSTEMS

***Junyi Wang**[*,1]**, Yi Weng**[2] **and Zhongqi Pan**[2]*
[1]Qualcomm Technologies Inc.
[2]Department of Electrical and Computer Engineering,
University of Louisiana at Lafayette

ABSTRACT

In optical fiber transmission systems, there has been an increasing interest in using digital signal processing (DSP) combined with coherent detection to increase the transmission distance, spectral efficiency (SE), etc. Nyquist pulse shaping can be used to improve the spectral efficiency (SE) by narrowing the signal spectrum without introducing inter-symbol interference in a wavelength-division-multiplexed (WDM) system. Such systems are usually called Nyquist-WDM systems. In this chapter, we introduce digital signal processing for Nyquist-WDM systems. Our discussion covers the following topics. (1) Introduction of Nyquist spectral shaping in optical fiber WDM systems. (2) Theoretical background of Digital Nyquist spectral shaping. (3) Transmitter-side DSP and digital-to-analog converter (DAC) for Nyquist spectral shaping. (4) Receiver-side Matched filter design for Nyquist-WDM systems.

---

[*] Corresponding author: E-mail: wjijackcy@gmail.com

Topic (1) is an introduction of Nyquist spectral shaping in optical fiber WDM systems.

Topic (2) introduces theoretical background about Nyquist spectral shaping and digital filters.

Topic (3) discusses about transmitter-side DSP and DAC. In topic (3), the required tap number or overlap length are investigated for finite-impulse-response (FIR) or frequency-domain-equalization (FDE) filters. A FDE filter with a fast Fourier transform (FFT) size of 64 and an overlap length of 8 points performs slightly better than a 17-tap FIR filter. 25 taps are enough for a FIR filter to do Nyquist spectral shaping for PDM-QPSK WDM systems, with negligible optical signal-to-noise ratio (OSNR) penalty. And 37 taps are sufficient for a digital FIR filter to generate Nyquist-WDM PDM-16QAM signals. In topic (3), minimum required DAC speed is also investigated for generating Nyquist-WDM signals. The electrical filter shape and spectral pre-emphasis have a big impact on the system performances. By using a 4th-order super-Gaussian electrical filter and spectral pre-emphasis, 1.2-samples/symbol sampling speed is enough for DACs to generate PDM-16QAM Nyquist-WDM signals at 1.05-symbol-rate channel spacing, with OSNR penalties less than 0.4 dB for 28-Gbaud PDM-16QAM. In topic (3), a 2-channel interleaved DAC is also investigated to generated Nyquist-WDM signals, with zero-order holding at the speed of the symbol rate. The required spectral pre-emphasis, bandwidth and resolution are investigated. With proper spectral pre-emphasis, both the required electrical filter bandwidth and DAC resolution are similar to those with a conventional DAC used. Details are described in the context for the required parameters. A polyphase FIR filter structure is demonstrated in the transmitter to increase the processing parallelism and to reduce the calculation by half.

Topic (4) investigates the receiver-side matched filter design for RRC spectrally shaped Nyquist-WDM Systems. The required tap number of the receiver-side adaptive equalizer can be greatly reduced by adding matched filtering in the static chromatic dispersion equalizer. The matched filter makes a Nyquist-WDM system become more sensitive to the frequency offset between the lasers at the transmitter and receiver. Frequency-offset monitoring and compensation can reduce the frequency offset to a tolerable range. With a roll-off factor of 0.1, the OSNR penalty at BER 10^{-3} can be less than 0.2 dB from 1 GHz frequency offset.

1. INTRODUCTION

There has been an prevalent trend in using advanced multilevel modulation formats combined with digital signal processing (DSP) to increase the optical fiber communication capacity or spectral efficiency (SE).

Orthogonal frequency-division multiplexing (OFDM) can achieve high SE by packing many subcarriers at symbol-rate spacing [1,2]. However, OFDM requires frequency locked subcarriers and subcarriers cannot be added or dropped independently. Another technique is to use single-carrier modulation with transmitter-side (Tx) tight filtering to reduce the channel spacings to be close to the symbol rate in wavelength-division-multiplexing (WDM) systems.

When tight filtering is used to reduce the channel spacing, there are both intra-channel and inter-channel impairments. One major intra-channel impairment is inter-symbol interference (ISI) caused by the limited channel bandwidth. The inter-channel impairment is from the crosstalk from neighboring channels due to non-ideal filtering. In high-SE WDM system design, one challenge is to minimize both intra-channel and inter-channel impairments. With sophisticated detection techniques such as maximum a posteriori (MAP) detection at the receiver-side (Rx) to mitigate ISI induced impairments, it is possible to reduce the channel spacing even narrower than the symbol rate [3-8]. However, MAP detection needs some complicated digital signal processing (DSP) for making decisions at the receiver, and also requires some overhead for the necessary pattern learning. Another approach is to perform Tx spectral shaping (or pulse shaping) according to the Nyquist criterion to generate ISI-free signals. This technique can reduce the channel spacing close to the symbol rate with no ISI or negligible ISI, and is referred to as Nyquist-WDM. The channel spacing can be even reduced to be less than the symbol rate in supper-Nyquist or faster-than-Nyquist-WDM systems [9-11]. Nyquist spectral shaping is ISI free in theory. The crosstalk between channels can also be negligible by proper design [4]. A lot of the demonstrated Nyquist-WDM systems used a specially designed Nyquist optical filter [5], or an ordinary optical filter plus a pulse carver [6] to generate Nyquist-WDM signals. An alternative way for generating Nyquist-WDM signals is to use DSP at the transmitter. Nyquist-WDM systems with digital FIR filters have been demonstrated in [8]. Both FIR filters in the time domain and FDE filters in the frequency domain can do the digital filtering [12-20].

The rest of this chapter is organized as follows. Section 2 introduces theoretical background about Nyquist spectral shaping and digital filters for Nyquist spectral shaping. Section 3 discusses about Tx side DSP and DAC, including the required FIR tap number and FDE overlap length, minimum required DAC speed aided by 4-order Gaussian analog filtering, and a interleaved DAC structure to enable digital Nyquist spectral shaping with zero-order holding at the symbol rate. Section 4 discusses about Rx matched filtering design to reduce blind equalizer complexity for Nyquist Spectrally shaped systems. Section 5 summarizes this chapter.

2. Theoretical Background

This section introduces some background knowledge about Nyquist spectral shaping and digital filters to achieve Nyquist spectral shaping.

2.1. Nyquist-Criterion Spectral Shaping

ISI-free operation can be achieved if the spectrum of a signal satisfies the Nyquist criterion [21]:

$$\sum_{m=-\infty}^{\infty} H(f + m/T) = T \tag{2.1}$$

where T is the signal symbol period. The smallest bandwidth of a signal with zero ISI is 1/T. Its spectrum is rectangular and the pulse shape is a sinc function:

$$h(t) = \sin c(\frac{\pi t}{T}) = \frac{\sin(\pi t / T)}{\pi t / T} \tag{2.2}$$

However, such signal's infinitely long impulse response makes it fundamentally impossible to achieve. Moreover, because the tails of h(t) decay as 1/t, a large number of taps are required to approximate this filter in practice. And a small timing error in sampling can cause a large number of ISI components.

A more practical and widely used spectrum which also satisfies the ISI-free Nyquist criterion is the raised cosine (RC) spectrum, expressed as follows.

$$H(f) = \begin{cases} T, & |f| \le \frac{1-\beta}{2T} \\ \frac{T}{2}\left\{1 + \cos[\frac{\pi T}{\beta}(|f| - \frac{1-\beta}{2T})]\right\}, & \frac{1-\beta}{2T} \le |f| \le \frac{1+\beta}{2T} \\ 0, & otherwise \end{cases} \quad (0 \le \beta \le 1) \tag{2.3}$$

where β is the roll-off factor. The corresponding pulse shape is as follows.

$$h(t) = \frac{\sin(\pi t / T)}{\pi t / T} \frac{\cos(\pi \beta t / T)}{1 - 4\pi \beta^2 t^2 / T^2} \tag{2.4}$$

Equation (2.4) shows that the tails of the signal decay as $1/t^3$ for $\beta > 0$. Consequently, the impulse response decays faster and a timing error induces much smaller ISI for the signal. Compared with the signal with rectangular spectrum, the signal with the RC spectrum occupies an extra bandwidth of β/T.

Note that the ISI-free Nyquist criterion includes both transmitter and receiver filter responses. In an optimal communication system, matched filters are used at the receiver. In this case, one can generate a signal with a RRC shaped spectrum at the transmitter, which is a square-root of (2.3). With the matched filter at the receiver, the overall response of the system is RC and ISI-free.

2.1. Digital Filters for Nyquist spectral shaping

Nyquist spectral shaping can be generated with digital finite-impulse-response (FIR) filters or frequency-domain-equalization (FDE) filters. The following context introduces the structures of a digital FIR filter and a digital FDE filter.

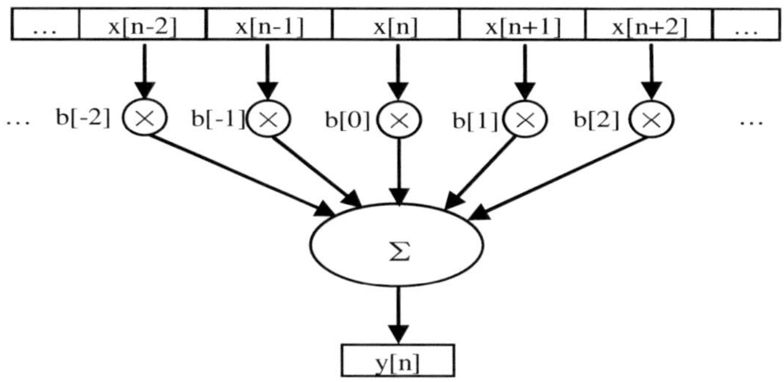

Figure 2.1. Digital FIR filter (x[n]: input digital signals. b[n]: weights. y[n]: output. Σ: sum.).

A digital FIR filter is a discrete-time FIR filter, whose output is a weighted sum of a center input and its neighbors, as shown in Figure 2.1. The input digital sequence, the weights, and the output sequence are represented by x[n], b[n], and y[n], respectively. In digital FIR filters, b[n] denotes the values of the impulse response at discrete points in time.

Figure 2.2 shows the principle of an overlap-and-save FDE filter. With an overlap-and-save FDE filter, the digital inputs are divided into blocks, such that the block length equals the size of the underlying fast Fourier transform (FFT) and there are some overlap samples between neighboring blocks. Each block is first converted to the frequency domain, and after the equalization in the frequency domain, the block is converted back to the time domain by inverse fast Fourier transform (IFFT). The overlap in a block is discarded when this block is stored as filtered results. To understand the principle/performance of an overlap-and-save FDE filter, one can use its equivalent FIR filter. That is, the performance of a FDE filter with an overlap length longer than its impulse response length (IRL) is the same as that of a FIR filter with a length longer than the IRL.

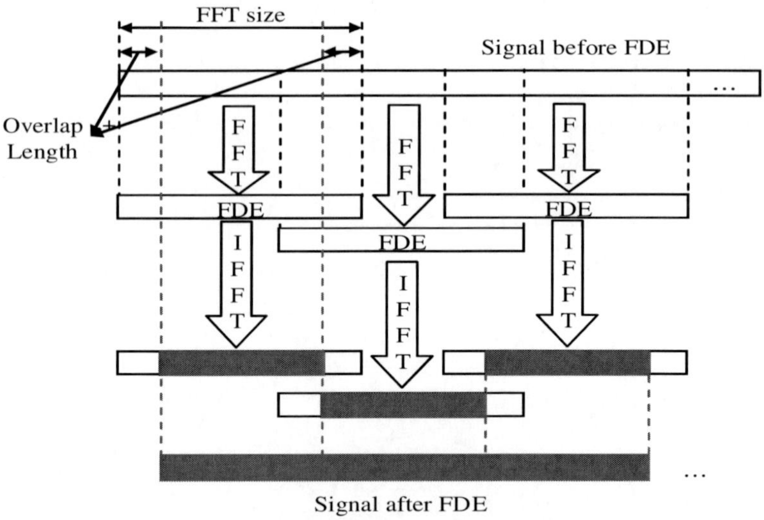

Figure 2.2. Overlap-and-save FDE filter (Blue parts represents the remaining samples).

The IRL can be defined in different ways. One is the length when the IR drops to some threshold (such as 1% of the centered maximum value), and another one is the length of a FIR filter that makes the normalized out-of-band power of the filtered signal to start converging (such as, the difference from the converged value by less than 0.1 dB can be regarded as"starting to converge"). The normalized out-of-band power is the power outside the band [$-R_S/2$, $R_S/2$] normalized to the total signal power. The IRLs of a digital Nyquist filter to generate RC and RRC spectral shapes with different roll-off factors are measured and shown in Tables 2.2.1 and 2.2.2. The IRLs under the two definitions can be different, especially when RRC spectral shaping is used. Since the out-of-band power can cause crosstalk between channels in a WDM system and is a more serious concern than ISI for RC or RRC spectral shaping [22,23], the second definition of the IRL should be used.

Figure 2.2.3 shows the normalized out-of-band power versus the FIR length or FDE overlap length, with RRC 0.1 spectral shaping, respectively. The input of the digital filters is 8192 symbols of digital BPSK signal, which is generated from the odd sequence of a pseudo random binary sequence (PRBS) with a pattern length of 2^{13}-1. The required FIR filter length is M_s symbols (it is $M=2\times M_s+1$ taps with oversampling of 2 samples per symbol), which is also the IRL under the second definition. The required FDE filter overlap length, defined as the minimum overlap length to get the normalized out-of-band power converge, is n_s symbols when the FFT size is big enough. In Figure 2.2.3, n_s is 6 symbols with RRC 0.1 spectral shaping.

Table 2.2.1. IRL when IR drops to 1% of its center value for a BPSK signal

Roll-off factor	0.05	0.1	0.15	0.2
RC IRL (symbol)	36	22	16	12
RRC IRL (symbol)	30	20	14	14

Table 2.2.2. IRL when the normalized out-of-band power starts to converge for a BPSK signal (for RC 0.1 it converges to -24.36 dB, for RRC 0.1 it converges to -17.42 dB)

Roll-off factor	0.05	0.1	0.15	0.2
RC IRL (symbol)	48	22	16	12
RRC IRL (symbol)	22	12	8	6

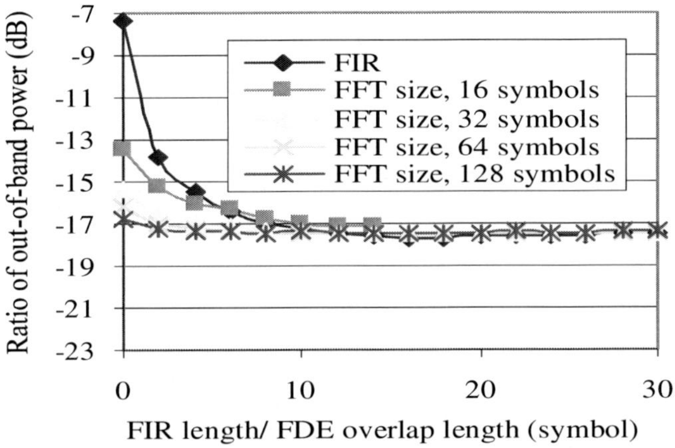

Figure 2.2.3. Normalized out-of-band power with RRC 0.1 spectrum shape.

Although one can figure out a minimum required FIR tap number or FDE overlap length by investigating the normalized out-of-band power, some attention need to be paid to. The channel spacing might vary in the transmission systems, which can be 1.1 times the symbol rate or another value. And some other factors (such as the modulation format, the amplified spontaneous emission noise, etc.) may affect the impact of the normalized out-of-band power. A simulation considering these factors are required to find out the actually required FIR tap number or FDE overlap length, a converged BER should be the criterion to determine the minimum required FIR tap number or FDE overlap length.

3. TRANSMITTER DESIGN

3.1. Generation of Spectrally Efficient Nyquist-WDM Signals Using **Digital Filters**

3.1.1. Generation of PDM-QPSK signals for Nyquist-WDM Systems

To investigate the crosstalk and ISI effects in a Nyquist-WDM system, a simple 3-channel WDM polarization-division-multiplexed (PDM) quadrature-phase-shift-keying (QPSK) system using MATLAB is simulated and the back-to-back performance is examined. The simulation setup is shown in Figure 3.1.1. In each channel a 112-Gbit/s PDM-QPSK signal is digitally pre-filtered to generate a RRC spectral shape. For one polarization in each channel, 8192 symbol digital QPSK signal are generated from a PRBS with pattern length 2^{13}-1. A digital filter is designed based on a sampling rate of 2 times the symbol rate. The nonlinearity of the MZM is compensated after the digital filter, by calculating the reverse function of the MZM modulation curve, which can be implemented by lookup tables in the hardware. To modulate the signal onto an optical carrier through the MZM, the signal is converted to the analog regime by a digital-to-analog converter (DAC). Although there are different designs for DAC [24], a straightforward and easy-to-implement design is chosen for our simulation. The DAC holds each digital value for half a symbol period and then filters it with a 4th-order super Gaussian (SG) electrical filter with a 3-dB bandwidth of 0.625 times the symbol rate R_S. The analog electrical signal is then used for PDM-QPSK modulation.

Three PDM-QPSK WDM channels are combined without any optical filtering. The system performance is quantified by the required optical-to-signal-noise ratio (OSNR) at a certain bit error rate (BER), and the OSNR is varied by changing amplified spontaneous emission (ASE) noise loaded before the receiver. At the receiver, a 4th-order SG electrical filter (e-filter) with a 3-dB bandwidth of $0.425R_S$ is used to extract the central channel for performance evaluation through DSP techniques in standard coherent receivers. A 3-dB bandwidth of $0.425R_S$ gives the best performance with the channel spacing of the symbol rate and the RRC 0.1 spectral shaping [23]. The random polarization rotation is considered, and a 25-tap butterfly FIR equalizer is used for polarization de-multiplexing and ISI compensation, adjusted via the CMA. The BER is then calculated by direct error counting. We tried different time delays and phase differences among channels [25], and the performance variation caused by different time delays and phases is small, within 0.2 dB. The results shown in the paper are the average results over 100 random delays and initial phases among channels. The effects of laser linewidth of the transmitter and local oscillator (LO), and the frequency offset between them are ignored for simplicity.

Results with channel spacings of 1 and 1.1 times the symbol rate are shown in Figure 3.1.2. The RRC spectral shaping with different roll-off factors causes big differences in performances, especially when the channel spacing is small. The maximum performance difference with different roll-off factors shown in Figure 3.1.2 is about 0.9 dB when the channel spacing is 28 GHz. By increasing the number of taps, the required OSNR also decreases and reaches a converged value. For a channel spacing of 1.1 times the symbol rate and a symbol rate of 28 GBaud, the required OSNR at BER 10^{-3} for a 17-tap FIR filter is < 13.8 dB with roll-off factors from 0.1 to 0.2.

Laser frequency drifts could induce significant penalties in a system with narrow channel spacings. For a system with a channel spacing of 1.1 times the symbol rate, ±5% laser frequency drift (±1.4 GHz for 28 Gbaud) could reduce the effective channel spacing to the symbol rate. For the system with a channel spacing of 1.1 times the symbol rate and RRC spectral shaping using FIR filters of 17 taps, the OSNR penalty (compared to 13.55 dB, the performance of a single channel system without Nyquist pulse shaping) at BER 10^{-3} is about 1.0 dB even with ± 1.4-GHz laser frequency drift when the roll-off factor is 0.05 or 0.1 [23]. A roll-off factor of 0.1 would be an optimum choice.

The system performance is then investigated for a 3-channel WDM system at different channel spacings and different overlap length for RRC FDE filters. Figure 3.1.3 shows the back-to-back performance of a 3-channel 112-Gbit/s Nyquist-WDM PDM-QPSK system using digital FDE filters. The FFT size is 64 with oversampling of 2.

With RRC spectral shaping, roll-off factors of 0.05 and 0.1 give better performances. With a channel spacing of 1.1 times the symbol rate and an overlap length of 8 points, the required OSNR at BER 10^{-3} is less than 13.7 dB when the roll-off factor is between 0.1 and 0.2. And the OSNR penalty (compared to 13.55 dB) at BER 10^{-3} is about 0.8 dB even with ± 1.4-GHz laser frequency drift when the roll-off factor is 0.05 or 0.1. So 0.1 would be an optimum choice for the roll-off factor with RRC spectral shaping. It can be noticed that this performance is slightly better than that with a 17-tap FIR filter.

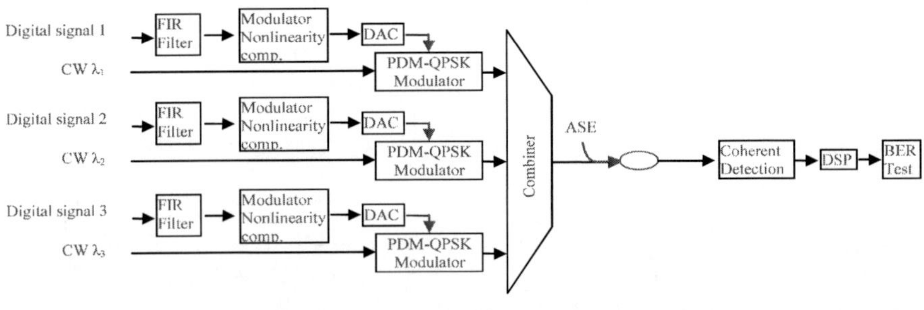

Figure 3.1.1. Simulation setup of a 3-channel Nyquist-WDM system.

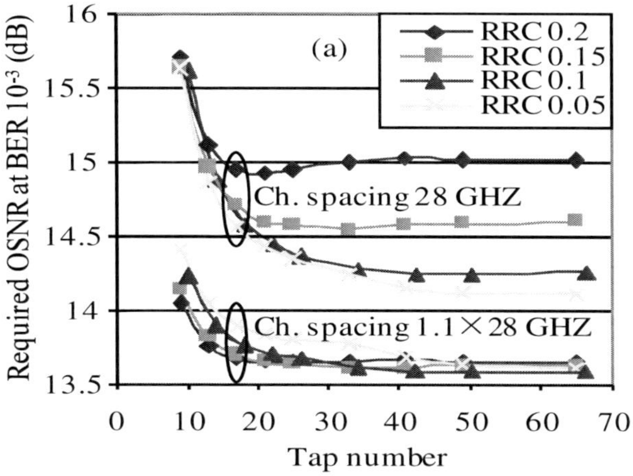

Figure 3.1.2. Performance of 112-Gb/s PDM-QPSK generated by FIR filters with different tap numbers for different RRC spectrum shapes. RRC 0.2: RRC spectrum shape with roll-off factor 0.2.

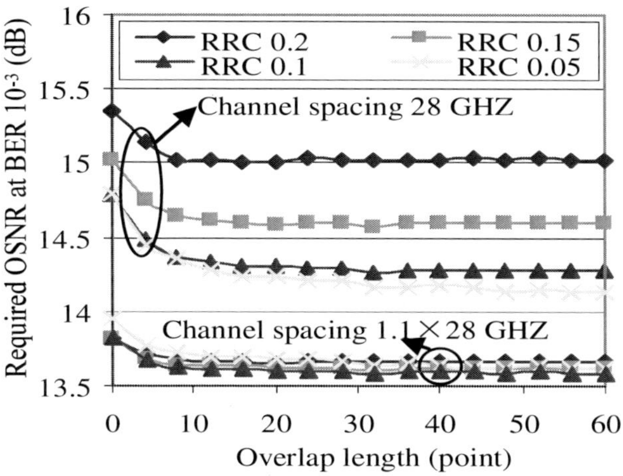

Figure 3.1.3. Performance of 112-Gb/s PDM-QPSK generated by FDE filters with FFT size of 64 taps and different overlap lengths for RRC spectrum shapes with different roll-off factors. RRC 0.2: RRC spectrum shape with roll-off factor 0.2.

The comparison of hardware complexity with FIR and FDE filters for Nyquist filter can go beyongd just comparing the required tap number and overlap length, as there are various designs to reduce the hardware complexity. Polyphase stuctrures are usually used in pulse shaping FIR filters to reduce the number of multiplications, while the optimization of hardware complexity with FDE filters can be very complicated. As a result, instead of FDE filter, FIR filters are usually used for Nyquist spectral shaping. Later discussion in this chapter uses FIR filter for Nyquist spectral shaping.

3.1.2. Generation of PDM-16QAM signals for Nyquist-WDM Systems

The simulation setup is almost the same as that in Figure 3.1.1, except that the modulation format is changed [26]. In our MATLAB simulations, a simple 3-channel polarization-division-multiplexed (PDM) 16-ary quadrature-amplitude-modulation (16QAM) Nyquist-WDM system is used to evaluate back-to-back performance with crosstalk and ISI impairments. In each channel a 224-Gbit/s PDM-16QAM signal is digitally pre-filtered to a RRC spectral shape. A 2^{13}-1 pseudorandom binary sequence (PRBS) is mapped to two 4-level signals with Gray coding for 16QAM generation, with one for inphase (I) and the other for quadrature (Q), respectively. The 4-level signals are upsampled to a rate of $2R_S$ with Dirac data pattern and fed into the digital FIR filters [27]. To correct the spectral roll-off from zero-order holding and the following E-filter, a pre-emphasis is added to the RRC transfer function of the digital filter. The nonlinearity of the Mach-Zehnder modulator (MZM) is compensated after the FIR filter, which can be implemented by a lookup table in the hardware. To modulate the signal onto an optical carrier through the MZM, the signal is converted to the analog domain by a DAC using zero-order hold. The DAC holds each digital value for a sampling period and then filters it with a SG or Gaussian E-filter. In our simulation, the speed of the DAC is set to $2R_S$ or lower than $2R_S$. For the DAC speed lower than $2R_S$, re-sampling is added before the DAC to make the DAC compatible with the previous DSP at the oversampling of 2. The I and Q analog electrical signals are then used to drive a MZM to generate a 16QAM signal. Two polarizations are generated by a relative delay and polarization rotation, and combined together with a polarization beam combiner (PBC). Three WDM channels are combined together without any optical filter, and random fractional delays and initial phase differences are added among the three PDM-16QAM WDM channels to obtain average bit error ratio (BER) performances with different crosstalk [25].

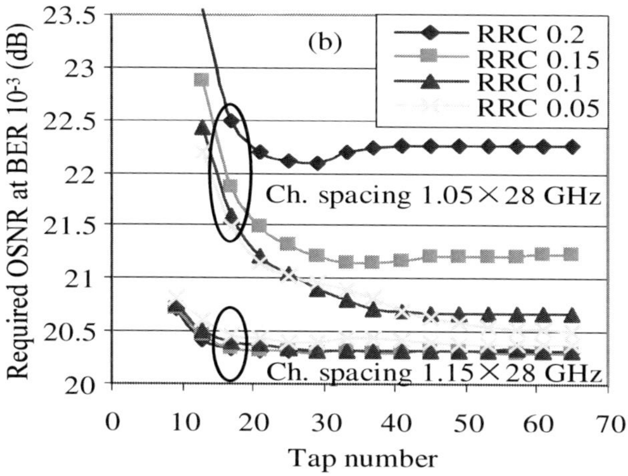

Figure 3.1.4. Required OSNR vs. tap number, with different roll-off factors and channel spacings at $1.05R_S$ and $1.15R_S$ (4th-order SG E-filter at Tx, BW $0.75R_S$).

We investigate the performance of a 3-channel PDM-16QAM WDM system at different channel spacings, with different RRC roll-off factors, and with different FIR filter tap numbers. The results with channel spacings of $1.05R_S$ and $1.15R_S$ are shown in Figure 3.1.4. It shows that a RRC spectral shape with a roll-off factor of 0.1 (RRC 0.1) is optimal considering the tradeoff between the performance and the filter tap number. RRC 0.05 only gives a slightly better performance at an expense of more filter taps with $1.15R_S$ channel spacing, and even gives a worse performance with $1.05R_S$ channel spacing due to higher requirements of receiver-side blind equalization. RRC 0.15 and RRC 0.2 give more OSNR penalties with channel spacing $1.05R_S$, because they have wider spectrum. A 37-tap FIR filter is sufficient for 16QAM Nyquist-WDM systems with the channel spacing of $1.05R_S$

3.2. Reducing the DAC Speed for Nyqist-WDM Systems

In principle, DSP can generate any signals with arbitrary shapes, but in reality, the complexity of DSP is limited by cost and power consumption. Therefore it is necessary to optimize the DSP to get the required performance at a minimum complexity. In optical fiber communication systems, typically 2 times oversampling digital-to-analog-converters (DACs) are used, and recently there were some studies on using lower speed DACs to generate signals. Reference [28] used a DAC speed of 1.5 times the symbol rate (R_S) to generate a raised-cosine (RC) spectrum with a roll-off factor of 0.5, but the spectrum is still too wide for a channel spacing close to R_S. In reference [29] a DAC speed of the symbol rate and spectral pre-emphasis were used to generate a RRC-like spectrum with a roll-off factor 0.15, but the spectrum does not satisfy the Nyquist criterion and is not narrow enough to reduce the channel spacing to $1.2R_S$.

In this section, using numerical simulations, the feasibility of DACs with speeds close to the symbol rate is investigated to generate 16QAM Nyquist-WDM signals. The results show that by using a 4th-order super Gaussian electrical filter (E-filter) and spectral pre-emphasis, DACs with $1.2R_S$ sampling speed can achieve a similar performance as those with $2R_S$ sampling speed.

3.2.1. Nyquist Spectral Shaping with DAC Speed Lower Than $2R_S$

According to Nyquist-Shannon sampling theorem, a function x(t) with no frequencies higher than B hertz (or lower than -B hertz) can be perfectly reconstructed by a series of sampling points with 1/(2B) seconds spaced [30]. A RRC signal with a roll-off factor of r has no frequencies higher than $(1+r)R_S/2$. In theory, an analog RRC signal can be reconstructed from a digital RRC signal at a sampling rate (f_S) $\geq (1+r)R_S$.

When zero-order hold as shown in Figure 3.2.1 is used for such reconstruction, harmonic frequency components are generated beyond $f_S/2$ as shown in Figure. 3.2.2. If not filtered out, these frequency components will generate crosstalk in WDM systems. An analog E-filter is usually used to filter out the undesired harmonic frequencies. When the sampling rate is close to $(1+r)R_S$, an E-filter with steep edges is preferred. For super-Gaussian (SG) E-filter, a higher order SG filter gives steeper edges as shown in Figure 3.2.3 and would be preferred when the DAC speed is close to $(1+r)R_S$.

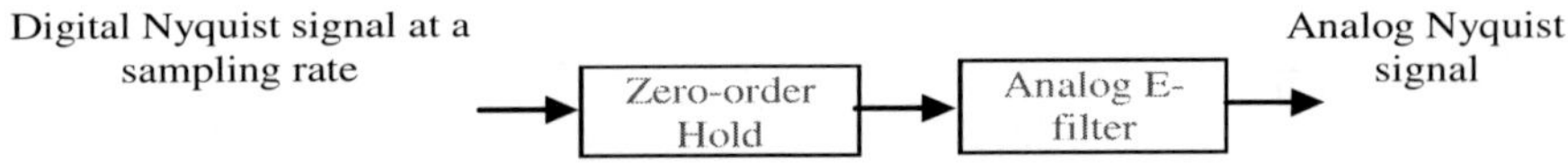

Figure 3.2.1. Digital Nyquist spectral shaping with zero-order holding for DAC.

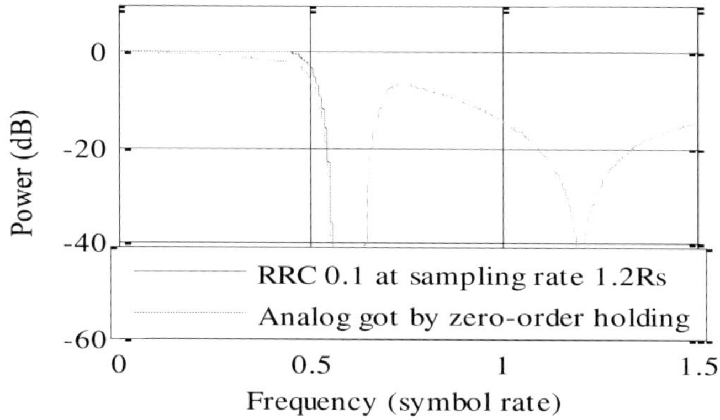

Figure 3.2.2. Analog spectrum obtained with zero-order holding from a digital RRC 0.1 signal at oversampling of 1.2.

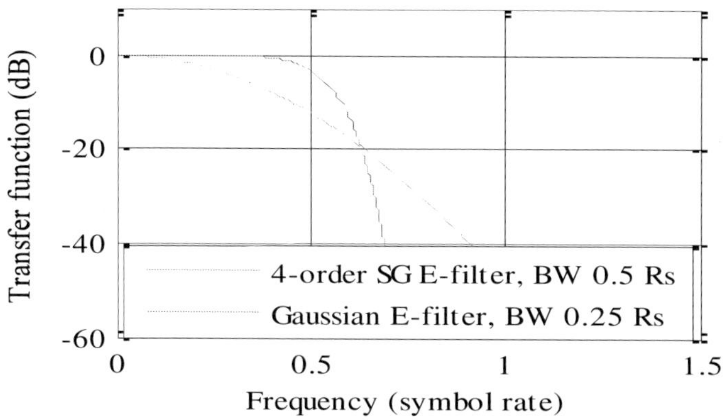

Figure 3.2.3. Transfer function of 4-order SG and Gaussian filter.

Besides non-steepness at the edges, the non-flatness of an E-filter below the frequency $(1+r)R_S/2$ also causes some spectral roll-off. A higher order SG filter gives more flatness in the low frequency range, and would be preferred when the DAC speed is close to $(1+r)R_S$. Another source of the roll-off is from the zero-order holding. The zero-order holding can be regarded as a convolution of Dirac-pattern inputs and a rectangular function [31]. In the frequency domain, it can be regarded as multiplication of the Fourier transforms of the input signal and the rectangular function. The Fourier transform of the rectangular function is a sinc function and gives roll-off to the input spectrum. The spectral roll-off can be pre-compensated by a spectral pre-emphasis in the DSP before zero-order holing.

3.2.2. Simulation Setup

In our MATLAB simulations, a simple 3-channel polarization-division-multiplexed (PDM)-16QAM Nyquist-WDM system is used to evaluate back-to-back performance with crosstalk and ISI impairments, as shown in Figure 3.2.4. In each channel a 224-Gbit/s PDM-16QAM signal digitally pre-filtered to a RRC spectral shape is generated. A 2^{13}-1 pseudorandom binary sequence (PRBS) is mapped to two 4-level signals with Gray coding for 16QAM generation, with one for inphase (I) and the other for quadrature (Q), respectively. The 4-level signals are upsampled to a rate of $2R_S$ with Dirac data pattern and fed into the digital FIR filters [27]. To correct the spectral roll-off from zero-order holding and the following E-filter, a pre-emphasis is added to the RRC transfer function of the digital filter. The nonlinearity of the Mach-Zehnder modulator (MZM) is compensated after the FIR filter, which can be implemented by a lookup table in the hardware. To modulate the signal onto an optical carrier through the MZM, the signal is converted to the analog domain by a DAC using zero-order hold. The DAC holds each digital value for a sampling period and then filters it with a SG or Gaussian E-filter. In our simulation, the speed of the DAC is set to $2R_S$ or lower than $2R_S$. For the DAC speed lower than $2R_S$, re-sampling is added before the DAC to make the DAC compatible with the previous DSP at the oversampling of 2. The I and Q analog electrical signals are then used to drive a MZM to generate a 16QAM signal. Two polarizations are generated by a relative delay and polarization rotation, and combined together with a polarization beam combiner (PBC). Three WDM channels are combined together without any optical filter, and random fractional delays and initial phase differences are added among the three PDM-16QAM WDM channels to obtain average bit error ratio (BER) performances with different crosstalk [25].

The system performance is then quantified by the required optical signal-to-noise ratio (OSNR) at a certain BER. The OSNR is varied by changing the amount of amplified spontaneous emission (ASE) at the receiver. At the receiver, a 4th-order SG E-filter with a 3-dB bandwidth (BW) of $0.45R_S$ is used to extract the central channel for the performance evaluation through DSP techniques in a PDM-16QAM coherent receiver. A 25-tap blind butterfly equalizer is adjusted first via the constant-modulus-algorithm (CMA) for pre-convergence, and then a radius-directed-equalization (RDE) algorithm is used to further optimize the equalizer for polarization de-multiplexing and ISI mitigation [32]. The BER is calculated by direct error counting, with 100 tests using a time window of 8192 symbols and for average performances from different ASE and crosstalk. The effects of the laser phase noise and the frequency offset between the transmitter and local oscillator (LO) are ignored for simplicity.

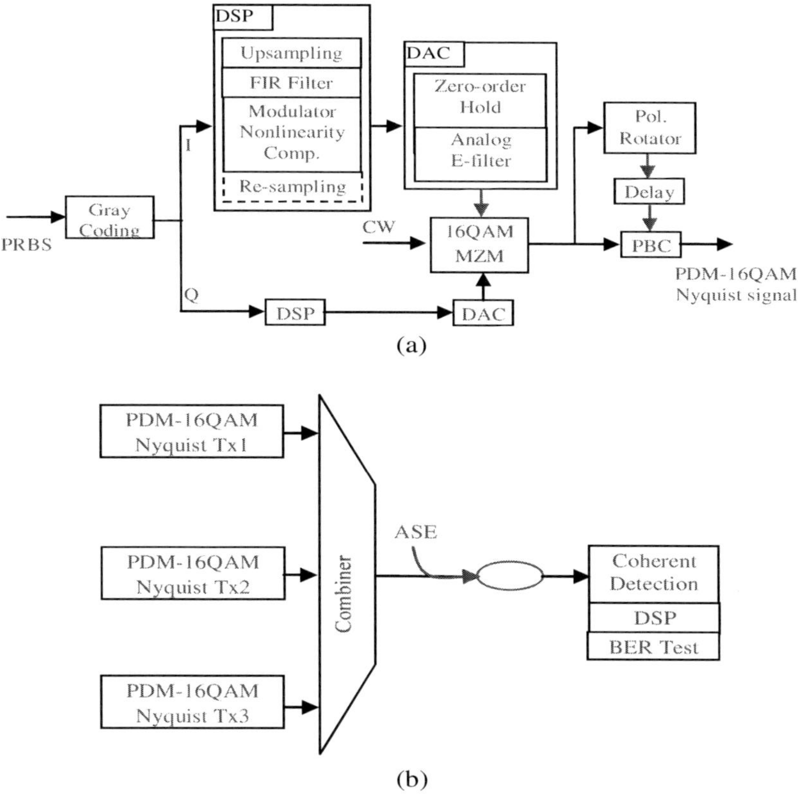

Figure 3.2.4. Simulation setup, (a) block diagram of a PDM-16QAM Nyquist transmitter, (b) a 3-channel PDM-16QAM Nyquist-WDM system.

3.2.3. Simulation Results and Discussion

Then the effects of lowering the DAC speed on the performance of the generated signals are investigated. DAC speeds lower than $2R_S$ are used to generate Nyquist 16QAM signals with a RRC 0.1 spectral shape. In order to make the DAC speed compatible with the previous DSP at an oversampling of 2, re-sampling is added before the DAC. The tap number of the FIR filter is fixed to 37 in following studies. Figure 3.2.5 shows the minimum required sampling rates for different SG E-filters, with the E-filter BW fixed at $0.5R_S$. As expected, for a higher order SG E-filter, the minimum DAC speed to have a negligible penalty is lower.

Figure 3.2.6 shows that even by choosing an optimum BW, the penalty with a 1st-order SG (or Gaussian) E-filter and spectral pre-emphasis is bigger than 1.5 dB when the DAC speed is $1.2R_S$. The penalty comes from the crosstalk due to non-steep edges of the E-filter, and insufficient FIR filter tap number due to the non-flatness of the E-filter. The main penalty is from the cross talk and the FIR tap number 37 is not sufficient for the RRC spectral shaping plus the spectral roll-off correcting, so the spectral pre-emphasis does not improve much on the performance. The crosstalk causes some obvious penalty even with a BW as small as $0.25R_S$, so it is not worthwhile to improve the performance by further reducing the BW and using a bigger filter tap number.

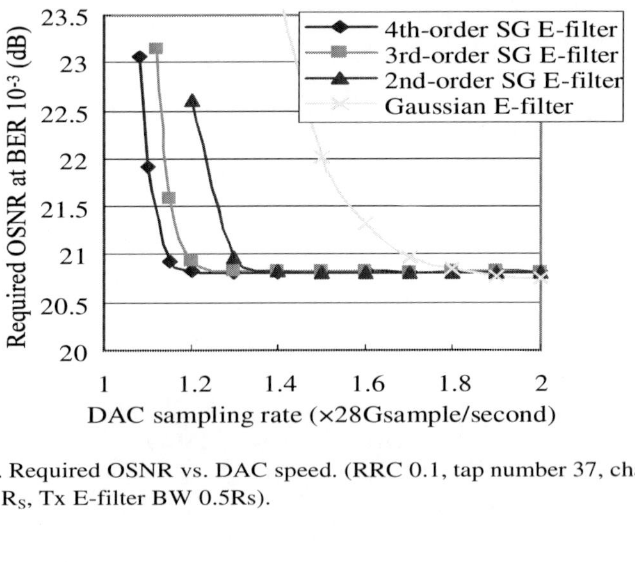

Figure 3.2.5. Required OSNR vs. DAC speed. (RRC 0.1, tap number 37, channel spacing $1.05R_S$, Tx E-filter BW 0.5Rs).

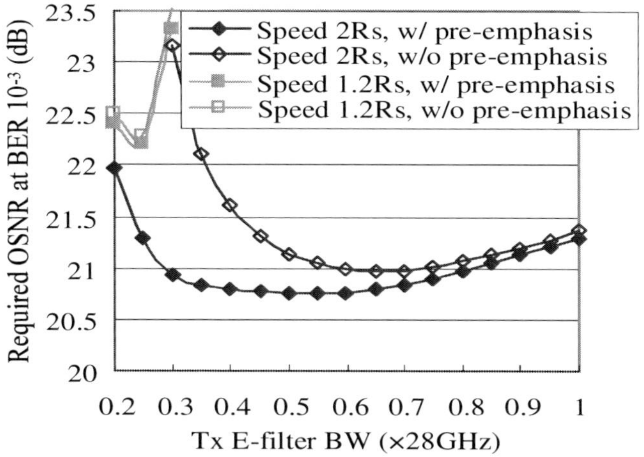

Figure 3.2.6. Required OSNR vs. Tx E-filter BW, with DAC speed $2R_S$ and $1.2R_S$. (RRC 0.1, tap number 37, channel spacing $1.05R_S$, Gaussian E-filter at Tx).

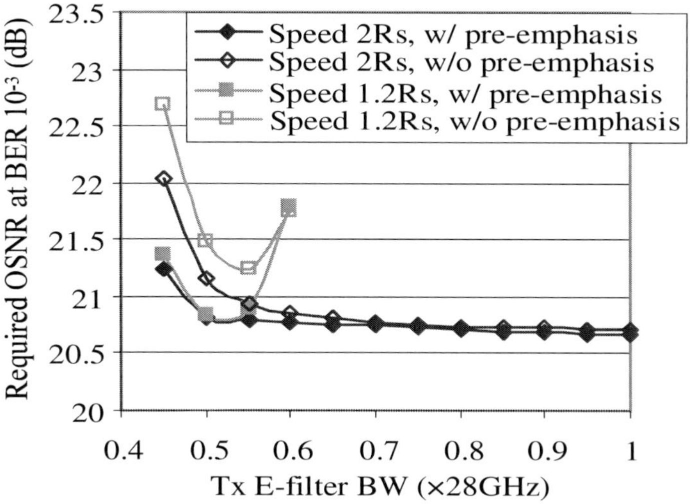

Figure 3.2.7. Required OSNR vs. Tx E-filter BW, with DAC speed $2R_S$ and $1.2R_S$. (RRC 0.1, tap number 37, channel spacing $1.05R_S$, 4th-order SG E-filter at Tx).

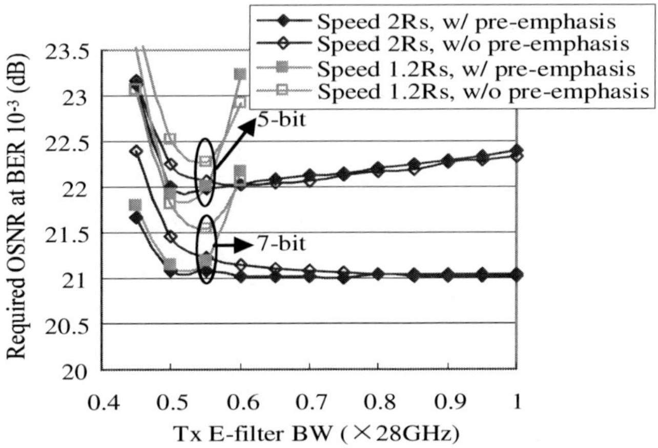

Figure 3.2.8. Required OSNR vs. Tx E-filter BW, with DAC speed $2R_S$ and $1.2R_S$. (RRC 0.1, tap number 37, channel spacing $1.05R_S$, 4th-order SG E-filter at Tx, DAC resolution 7 bits).

Figure 3.2.7 shows that by choosing an optimum BW $0.55R_S$, the penalty with a 4th-order SG E-filter is just 0.4-0.5 dB without spectral pre-emphasis when the DAC speed is $1.2R_S$. The spectral pre-emphasis can reduce the penalty to be negligible, with an optimum BW 0.5Rs.

In Figure 3.2.8, the DAC resolution is considered with the resolution of 7 bits and 5 bits. The conclusions from Figure 3.2.7 hold true with a 7-bit DAC resolution, and the 7-bit DAC resolution gives about 0.3 dB penalty. Although the penalty with a 5-bit DAC resolution is much bigger, the optimum performance with a $1.2R_S$ DAC speed is still about the same as that with a $2R_S$ DAC speed. When a 5-bit DAC resolution and a $2R_S$ DAC speed are used, the penalty increases slightly with a wider Tx E-filter BW, which is due to the crosstalk from the inaccurate spectral shaping.

3.3. Nyquist Spectral Shaping with Zero-Order Holding at the Symbol Rate

A RRC signal with a roll-off factor of r has all frequencies components within $(1+r)R_S/2$. According to the Nyquist-Shannon sampling theorem [30], the minimum sampling rate is $(1+r)R_S$ to sample and recover such signal. In later discussion we only consider integer oversampling ratio, to simplify the DSP design before the DAC. The minimum integer oversampling ratio for a digital RRC signal to perfectly recover an analog RRC signal is 1 if r=0, and is 2 if r>0.

Zero-order holding is an easy and straightforward way to do digital-to-analog conversion. Either a multiple-channel interleaved structure or a conventional 1-channel structure can be used, when zero-order holding is used for the digital-to-analog conversion. In a DAC with an interleaved structure, digital signals are split into multiple channels, and the digital signal in each channel is converted to an analog signal separately. The analog signals from all the channels are combined together to generate the final DAC output. For example, to generate a Nyquist signal with a roll-off factor of 0, some possible combinations of the digital signal sampling rate and the zero-order holding speed are shown in the following table.

The mathematical model of zero-order holding can be regarded as a convolution of a rectangular function and a Dirac-delta-pattern input signal [31]. Equation (3.3.1) and (3.3.2) show the definition of a rectangular function and its Fourier transform (FT) of a sinc function.

$$rect(t) = \begin{cases} 0 & if\ |t| < 0.5 \\ 0.5 & if\ |t| = 0.5 \\ 1 & if\ |t| > 0.5 \end{cases} \tag{3.3.1}$$

$$\int_{-\infty}^{\infty} rect(t) \cdot e^{-i2\pi ft} dt = \frac{\sin(\pi f)}{\pi f} = \sin c(f) \tag{3.3.2}$$

Equation (3.3.3) and (3.3.4) express the Dirac-delta-pattern input signals A(t) and B(t) from digital signals A(n) and B(m) at a sampling rate of $2R_S$ and R_S respectively, in which t_S denotes the symbol duration.

$$A(t) = \sum_{n=0}^{\infty} A(n) \cdot \delta(t - n \cdot \frac{t_S}{2}) \tag{3.3.3}$$

$$B(t) = \sum_{m=0}^{\infty} B(m) \cdot \delta(t - m \cdot t_S) \tag{3.3.4}$$

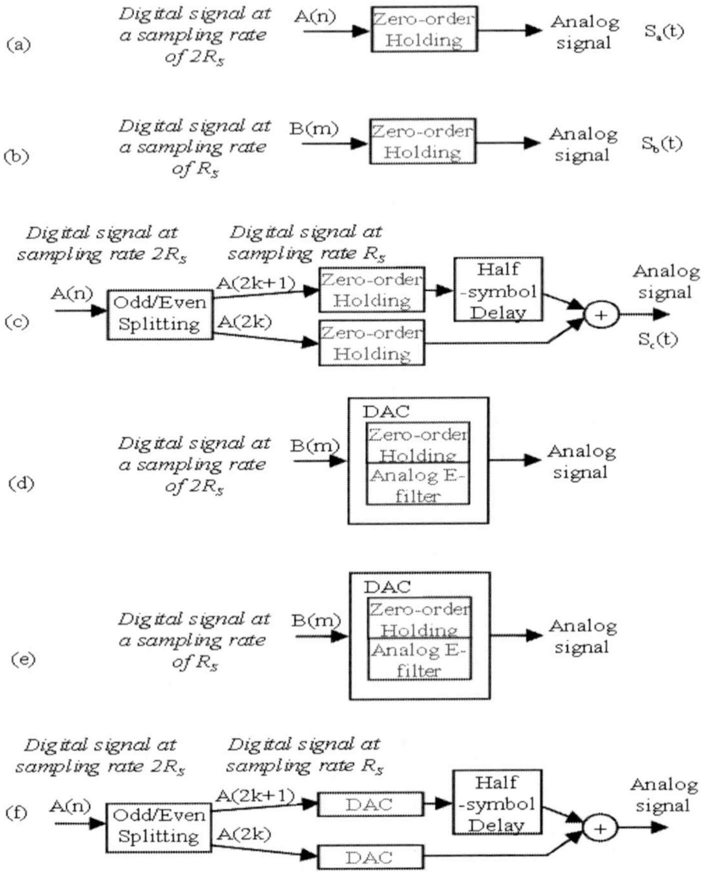

Figure 3.3.1. DACs by zero-order holding. (a-c), models without analog E-filters. (d-f) models with analog E-filters. (a, d), a conventional (or non-interleaved) DAC using zero-order holding at 2 times the symbol rate. (b, e), a non-interleaved DAC using zero-order holding at the symbol rate. (c, f), a 2-channel interleaved DAC using zero-order holding at the symbol rate.

Equation (3.3.5), (3.3.6) and (3.3.7) express the analog signals S_a, S_b and S_c generated from A(t), B(t) and A(t) at zero-order holding speed of $2R_S$, R_S and R_S respectively, with * denoting convolution.

Table 3.3.1. combinations of digital signal sampling rate and the zero-order holding speed for generating Nyquist signals with a roll-off factor of 0

	digital signal sampling rate	**Zero-order holding speed**	**Zero-order holding structure**
Comb. (a)	$2R_S$	$2R_S$	Non-interleaved as in Figure 3.3.1 (a, d)
Comb. (b)	R_S	R_S	Non-interleaved as in Figure 3.3.1 (b, e)
Comb. (c)	$2R_S$	R_S	Interleaved as in Figure 3.3.1 (c, f)

$$S_a(t) = \left[\sum_{n=0}^{\infty} A(n) \cdot \delta(t - n \cdot \frac{t_S}{2})\right] * rect(\frac{t}{t_S/2}) \quad (3.3.5)$$

$$S_b(t) = \left[\sum_{m=0}^{\infty} B(m) \cdot \delta(t - m \cdot t_S)\right] * rect(\frac{t}{t_S}) \quad (3.3.6)$$

$$S_c(t) = \left[\sum_{n=0}^{\infty} A(n) \cdot \delta(t - n \cdot t_S)\right] * rect(\frac{t}{t_S}) \quad (3.3.7)$$

Equation (3.3.8), (3.3.9) and (3.3.10) show that the convolutions can be regarded as multiplications in the frequency domain. Therefore, zero-order holding can be regarded as translating a digital signal into an analog Dirac-delta-pattern signal and then multiplying its Fourier transform with a sinc function in the frequency domain, which generates spectral roll-off and harmonic frequency components.

$$FT(S_a(t)) = \sin c(f/(2R_S)) \cdot FT(A(t)) \quad (3.3.8)$$

$$FT(S_b(t)) = \sin c(f/R_S) \cdot FT(B(t)) \quad (3.3.9)$$

$$FT(S_c(t)) = \sin c(f/R_S) \cdot FT(A(t)) \quad (3.3.10)$$

A(n) and B(m) represent digital RRC signals with a roll-off factor of 0. A(t) and B(t) can be calculated out as Dirac-delta-pattern signals based on

A(n) and B(m) according to Equation (3.3.3) and (3.3.4). In Figure 3.3.2, (a) and (b) show the spectrum of A(t) and B(t), while (c) and (d) show the transfer functions of sinc(f/R_S) and sinc($f/2R_S$).

In order to reduce crosstalk in WDM systems, undesired harmonic frequencies need to be filtered out or suppressed. Combination (b) in Table 3.3.1 has a DAC input of B(m) or B(t), and there is no gap from the main spectrum to the harmonic frequency components. Therefore, combination (b) has a strict or special requirement on the following analog E-filter [33]. This topic is not covered in later discussion.

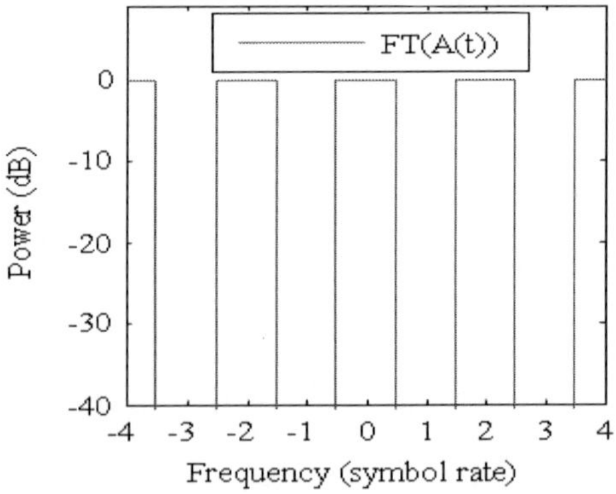

(a)

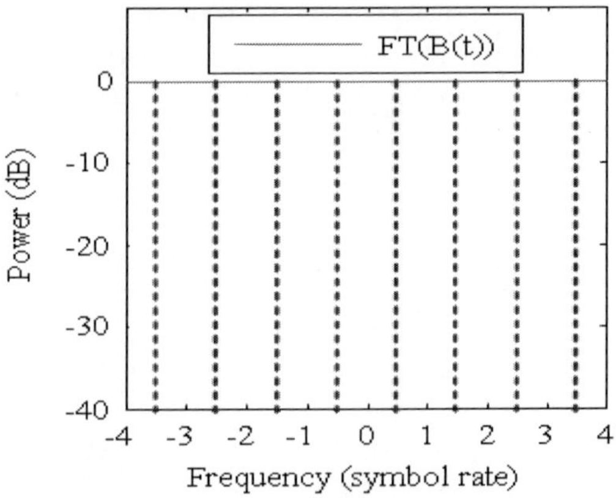

(b)

Figure 3.3.2. (Continued).

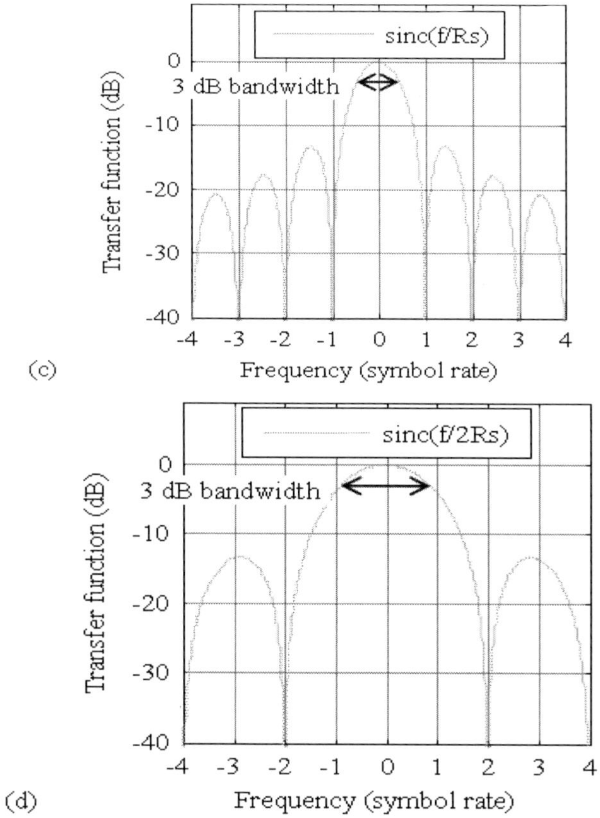

Figure 3.3.2. (a) Spectrum of A(t), (b) spectrum of B(t), (c) transfer function of sinc(f/R_S), (d) transfer function of sinc(f/2R_S).

Combination (a) and (c) in Table 3.3.1 use a DAC input of A(n) or A(t), in whose spectrum there are gaps from the main spectrum to the harmonic frequency components. With a low-pass analog E-filter, harmonic frequency components can be filtered out. Both the zero-order holding and the E-filter cause some spectral roll-off. The spectral roll-off from the analog E-filter is because of the following facts, the filter is not flat below the frequency (1+r)R_S/2 and the cutoff cannot be perfectly steep. Such spectral roll-off is the same for combination (a) and (c). From Figure 3.3.2 (c) and (d), the 3-dB bandwidth of zero-order holding at the symbol rate is much narrower than that of zero-order holding at 2 times the symbol rate. Therefore, compared to a conventional non-interleaved DAC, a 2-channel interleaved DAC generates a steeper spectral roll-off from zero-order holding. Fortunately, no matter either combination (a) or (c) is used, the spectral roll-off from the zero-order holding and the following analog E-filter can be pre-compensated, by a spectral pre-emphasis in the DSP before the zero-order holding. And the impairment from timing offset inaccuracy between the interleaved channels is also tolerable [34]. Combination (c) uses a zero-order holding at the symbol rate, half of that used combination (a), thus has an unique advantage in high speed Nyquist-WDM systems where the signal symbol rate can be 28 Gbaud or higher. Later discussions focus on combination (c), with a comparison to combination (a).

3.3.1. Polyphase FIR Filter Structure Combined with Interleaved DAC

Dirac-pattern signals are usually used for the FIR filter input [34]. Dirac-pattern signals can be got by upsampling signals at the symbol rate to a sampling rate of N times the symbol rate, with N as the upsampling ratio. Specifically, the upsampling can be regarded as setting the first sample to be the signal at that symbol and setting the other samples to be zero. According to [26], a polyphase FIR filter can be used to remove unnecessary multiplications with zeros when a Dirac-pattern signal passes a FIR filter. The multiplication number in a normal FIR filter is that in a polyphase FIR filter multiplied by the oversampling ratio. Here an oversampling ratio of 2 is used. A polyphase FIR filter can be naturally combined with an interleaved DAC. In Figure 3.3.3, the weights of a 5-tap (5-tap is just an example, the structure applies to an arbitrary tap number) polyphase FIR filter are split into to 2 groups, an even-indexed weights and an odd-indexed weights. A sub FIR filter with the even-indexed weights gives the even-indexed output, while another sub FIR filter with the odd-indexed weights gives the odd-indexed output. In a non-interleaved conventional DAC, a parallel-to-serial converter is necessary to combine the even-indexed and odd-indexed output to be a whole signal at an oversampling of 2 before the DAC. Such parallel-to-serial converter is not required in a 2-channel interleaved DAC. Either of the even-indexed and odd-indexed output from the FIR filter is put into a channel of the interleaved DAC directly. For an oversampling ratio of 2, compared to a non-polyphase FIR filter a polyphase FIR filter saves the multiplication number by half, and can save a high-speed parallel-to-serial converter if combined with a 2-channel interleaved DAC. Moreover, by using separate copies of the input signal, the sub filter with even-indexed weights and the sub filter with odd-indexed weights can be working in parallel at a lower computation speed than a non-polyphase FIR filter. In this way, this combination not only reduces the hardware complexity but also increases the processing parallelism.

3.3.2. Simulation Setup

With MATLAB, we simulated a 3-channel polarization-division-multiplexed (PDM) Nyquist-WDM system to evaluate back-to-back performance with crosstalk and ISI impairments, similar as shown in Figure 3.2.4. The modulation format is PDM-QPSK and PDM-16QAM. In each channel, a 112-Gbit/s PDM-QPSK or 224-Gbit/s PDM-16QAM signal is generated with RRC 0.1 spectral shaping. Firstly a 2^{13}-1 pseudorandom binary sequence is mapped with Gray coding to two 4-level signals for 16QAM generation or two 2-level signals for QPSK generation, with one for inphase (I) and the other for quadrature (Q), respectively. Polyphase digital FIR filters are then used for the spectral shaping and pre-emphasis. The transfer function of the FIR filters is a RRC function with a roll-off factor of 0.1 and some spectral pre-emphasis. A large enough tap number is chosen to be 37 for all simulations [23,26]. After the FIR filters the nonlinearity of the Mach-Zehnder modulator (MZM) is compensated, which can be implemented by a lookup table in the hardware. To modulate the signal onto an optical carrier through the MZM, the signal is converted to the analog domain by a DAC using zero-order holding followed by Gaussian analog E-filtering. In the simulation the DAC is either a conventional DAC at a speed of $2R_S$ or a 2-channel interleaved DAC at a speed of R_S. Pre-emphasis is added to the transfer function of the digital FIR filter to correct the spectral roll-off from zero-order holding and the analog E-filtering. The I and Q analog electrical signals drive a MZM to generate a QPSK/16QAM signal. Two polarizations are generated by a relative delay and polarization rotation, and combined together with a polarization beam combiner. Three WDM channels are combined together at a channel spacing of $1.1R_S$ without any optical filter, and
random fractional delays and initial phase differences are added among the three PDM-16QAM WDM channels to obtain average BER performances with different crosstalk [25].

The system performance is then quantified with the required optical signal-to-noise ratio (OSNR) at a BER of 10^{-3}. The OSNR is varied by changing the amount of amplified spontaneous emission (ASE) before the receiver. The signal is coherently detected at the receiver, and a 4th-order supper Gaussian analog E-filter with a 3-dB bandwidth (BW) of $0.6R_S$ is used to extract the central channel.

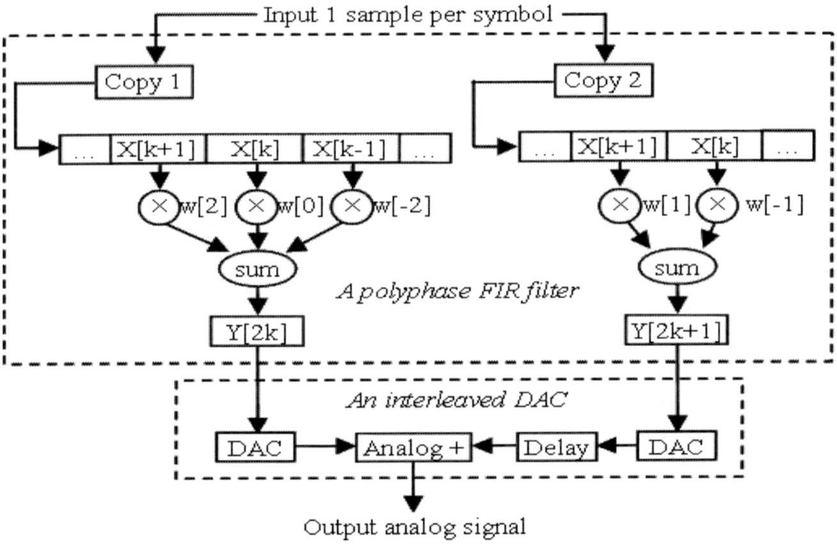

Figure 3.3.3. Combination of a 2-channel interleaved DAC and a polyphase FIR filter at an oversampling ratio of 2.

The extracted analog electrical signal is then sampled at rate of $2R_S$. Digital matched filtering is performed in frequency domain [35,36]. Then a FIR filter based blind equalizer is used for polarization de-multiplexing. With ideal Tx spectral shaping and Rx matched filtering, no residual ISI compensation is required. However, in a real implementation a non-ideal spectral pre-emphasis or DAC resolution may cause the imperfection of the Tx spectral shaping, and result in some Rx residual ISI. The residual ISI can be reduced by using an amount of tap number for the blind FIR equalizer. In the following discussion, the blind equalizer tap number is set as 25 for all simulations. The blind equalizer is adjusted via the constant-modulus algorithm (CMA) for PDM-QPSK signals. For PDM-16QAM signals, the blind equalizer is adjusted first via the CMA for pre-convergence and then a radius-directed-equalization algorithm for final convergence [32]. BER is calculated by direct error counting, with a time window of 8192 symbols and 100 tests for average performances with random ASE noise. The effects of the Tx and Rx laser phase noise and frequency offset are ignored, as their penalties can be compensated or reduced by some DSP algorithms and are independent from the Tx Nyquist spectral shaping.

3.3.3. Simulation Results and Discussion

We firstly investigate how the spectral pre-emphasis, the Tx E-filter bandwidth, and the DAC resolution affect a 112 Gbit/s RRC 0.1 Nyquist-WDM PDM-QPSK system at a channel spacing of 1.1R_S. Figure 3.3.4 shows the required OSNR decreases with the DAC resolution increasing, and finally converges. After the convergence, the spectral pre-emphasis improved (or reduced) the required OSNR by about 0.7 dB for a 2-channel interleaved DAC, while the improvement is very small for a conventional DAC, about 0.2 dB. Also from Figure 3.3.4, the required DAC resolution is 6 bits to have a required OSNR penalty below 0.1 dB.

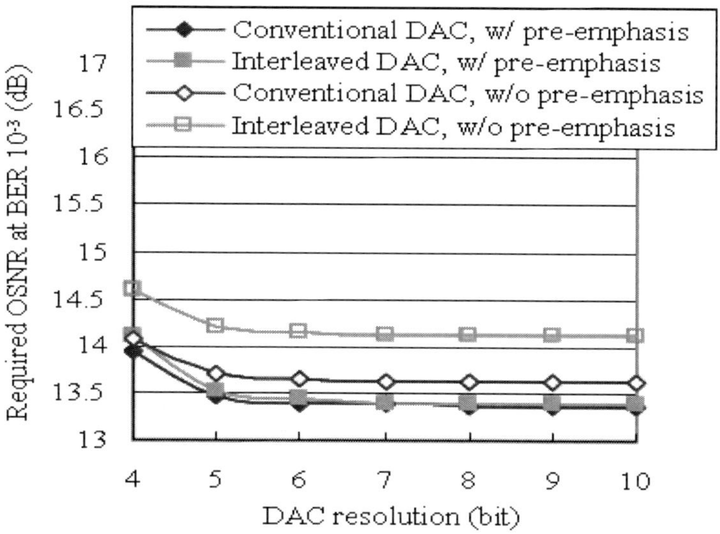

Figure 3.3.4. Required OSNR vs. DAC resolution for a 3-Channel PDM-QPSK Nyquist system, using conventional DAC and interleaved DAC, with and without spectral pre-emphasis (Tx Gaussian E-filter BW of half the symbol rate).

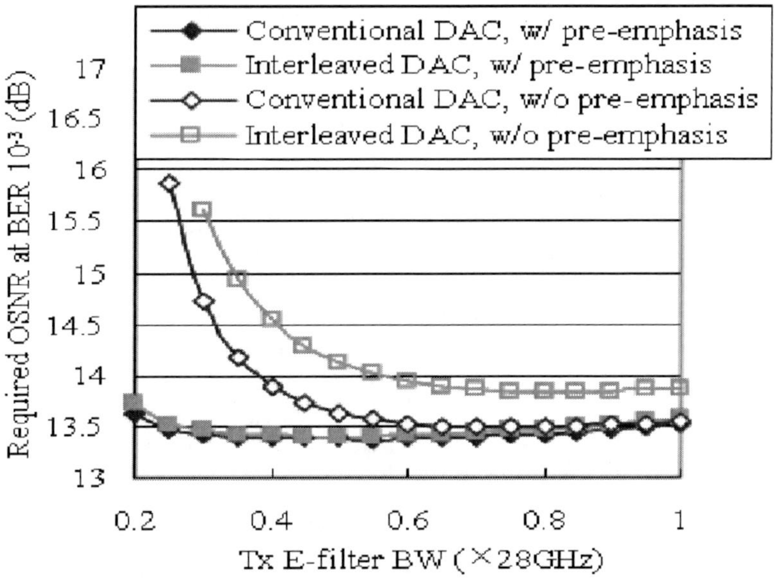

Figure 3.3.5. Required OSNR vs. Tx Gaussian E-filter bandwidth for a 3-Channel PDM-QPSK Nyquist system, using a conventional DAC and a 2-channel interleaved DAC, with and without spectral pre-emphasis (DAC resolution of 7 bit).

Figure 3.3.5 shows the required OSNR changes with the Tx analog Gaussian E-filter bandwidth. Similar as Figure 3.3.4, Figure 3.3.5 shows the penalty without the spectral pre-emphasis is much bigger for an interleaved DAC than a conventional DAC. This is because the extra spectral roll-off from a lower zero-order holding speed makes the spectral shaping drift more from the matched filtering design [36]. The Tx E-filter BW need be properly chosen. If the E-filter BW is too wide, a large amount of the harmonic frequency components from zero-order holding would remain after E-filter. If the E-filter BW is too narrow, more taps of the FIR filter would be required to achieve the spectral pre-emphasis. To have a required OSNR penalty below 0.1 dB, the Tx Gaussian E-filter bandwidth should be between $0.3R_S$ and $0.8R_S$. Both Figure 3.3.4 and Figure 3.3.5 show that with a proper spectral pre-emphasis, the performance, the required E-filter bandwidth and the required resolution of a 2-channel interleaved DAC are similar as those of a conventional DAC.

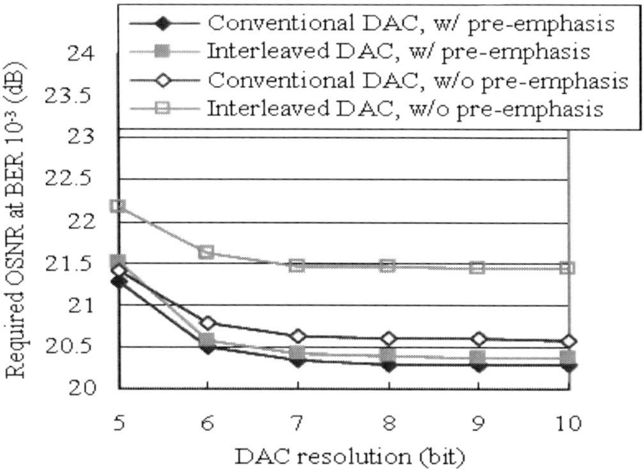

Figure 3.3.6. Required OSNR vs. DAC resolution for a 3-Channel PDM-16QAM Nyquist system, using a conventional DAC and a 2-channel interleaved DAC, with and without spectral pre-emphasis (Tx Gaussian E-filter BW of half the symbol rate).

Then we investigate how the spectral pre-emphasis, the Tx Gaussian E-filter bandwidth, and the DAC resolution affect a 224 Gbit/s Nyquist-WDM PDM-16QAM system, with a same spectral shaping of RRC 0.1 and a same channel spacing of $1.1R_S$. Similar as Figure 3.3.4, Figure 3.3.6 shows the required OSNR decreases with the DAC resolution increasing, and finally converges. After the convergence, the spectral pre-emphasis improved the required OSNR by more than 1 dB for a 2-channel interleaved DAC, while the improvement is very small for a conventional DAC, about 0.3 dB. Also from Figure 3.3.6, the required DAC resolution is 7 bits to have a required OSNR penalty below 0.1 dB.

Figure 3.3.7 shows the required OSNR changes with the Tx analog Gaussian E-filter bandwidth. Similar as Figure 3.3.5, Figure 3.3.7 shows the penalty without the spectral pre-emphasis is much bigger for an interleaved DAC than a conventional DAC. To have a required OSNR penalty below 0.1 dB, the required Tx Gaussian E-filter bandwidth should be between $0.35R_S$ and $0.65R_S$. Both Figure 3.3.6 and Figure 3.3.7 show that with a proper spectral pre-emphasis, the performance, the required E-filter bandwidth and the required resolution of a 2-channel interleaved DAC are similar as those of a conventional DAC.

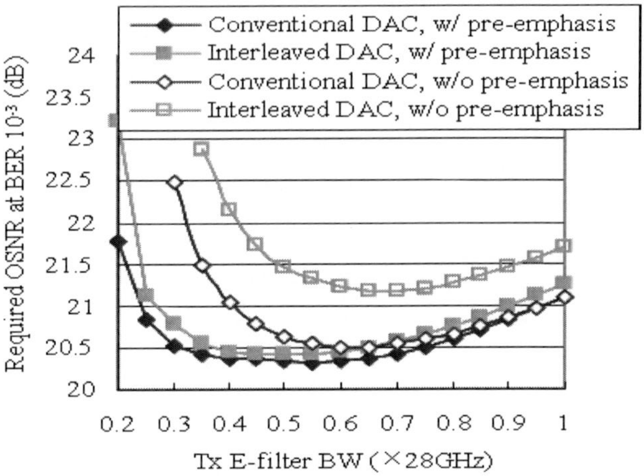

Figure 3.3.7. Required OSNR vs. Tx Gaussian E-filter bandwidth for a 3-Channel PDM-16QAM Nyquist system, using conventional DAC and 2-channel interleaved DAC, with and without spectral pre-emphasis (DAC resolution of 7 bit).

4. Receiver Design

In Section 3, it can be noticed that RRC 0.05 has a more strict requirement on the Rx blind equalization than RRC 0.1. Generally speaking, for a Nyquist signal with a smaller roll-off factor, a larger number of taps and a longer convergence time are required for a FIR filter based BE. To reduce the complexity of the BE, we propose to incorporate a matched filter in the bulk CD equalizer [35,36]. Simulation results of PDM-QPSK and PDM-16QAM Nyquist systems with RRC spectral shapes show that adding a RRC matched filter in the bulk CD equalizer significantly reduces the complexity of the BE. It's also found that including the matched filter function does not add additional complexity to the CD equalizer.

4.1. Simulation Setup

In the MATLAB simulation, a single-channel PDM-QPSK and PDM-16QAM system are used to investigate the performance and complexity of the

BE at the receiver in both a back-to-back (BTB) system and a transmission system with 1000 km uncompensated standard single-mode-fiber (SSMF). Note that only CD is considered in the 1000-km transmission system. The simulation setup is illustrated in Figure 4.1. For comparison, a regular non-return-to-zero (NRZ) PDM-QPSK/16QAM transmitter and a Nyquist PDM-QPSK/16QAM transmitter are used. The 112-Gbit/s PDM-QPSK and 224-Gbit/s PDM-16QAM signals are generated from a 2^{13}-1 pseudo-random bit sequence (PRBS). Nyquist spectral shaping is realized through DSP-based digital Nyquist filtering with a sampling rate of 2 times the symbol rate (R_S). An 81-tap FIR filter is used for Tx RRC spectral shaping. Dirac data pattern is used before the FIR filter and spectral pre-emphasis is also used to correct the spectral roll-off due to the following zero-order hold and .non-flat E-filter. The E-filter bandwidth is 21 GHz for all the setups. The system performance is quantified by the required optical signal-to-noise ratio (OSNR) at bit error ratio (BER) of 10^{-3}. The OSNR is varied by changing ASE noise loaded to the receiver. At the receiver, a 4-th order supper Gaussian electrical filter with a 3-dB bandwidth of $0.6R_S$ is used before digital sampling at a sampling rate of $2R_S$. CD compensation is performed in a static overlap-and-save FDE. Then a FIR filter based BE is used for polarization de-multiplexing and residual ISI compensation. The BE is adjusted via the constant-modulus algorithm (CMA) for PDM-QPSK signals. For PDM-16QAM signals, the BE is adjusted first via the CMA for pre-convergence and then a RDE algorithm is used to further optimize the equalizer. BER is calculated by direct error counting, with a time window of 8192 symbols and 100 tests for average performances. The effects of the laser phase noise are ignored. The frequency offset between the transmitter and LO is also ignored, assuming that the frequency offset can be monitored and the center frequency of the matched filter can be shifted to the offset.

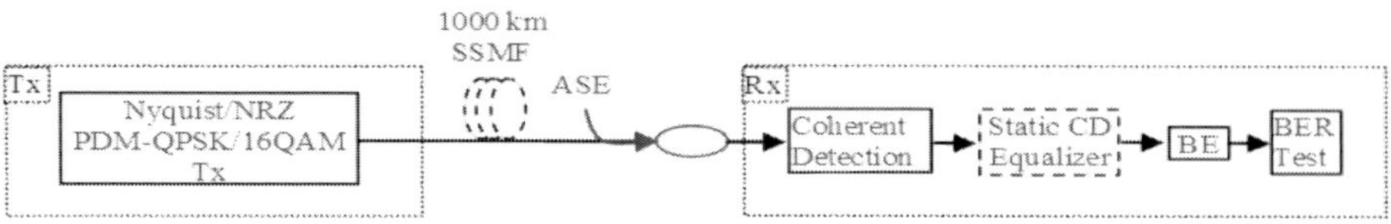

Figure 4.1. Simulation setup of a single-channel Nyquist spectrally shaped or NRZ system.

4.2. Simulation Results and Discussion

Firstly the BE complexity without and with including the matched filter in the CD equalizer for PDM-QPSK is investigated in a BTB system with Tx RRC 0.05 spectral shaping. The result is shown in Figure 4.2, which depicts the required OSNR versus the number of BE taps for the two receiver configurations. For comparison, a regular NRZ PDM-QPSK signal without using a matched filter is also plotted in the figure. The minimum required BE tap number is about 15 to have an OSNR penalty below 0.1 dB when no matched filter is included in the CD equalizer, and can be reduced to 3 if a matched filter is added in the CD equalizer. As a reference, a regular NRZ PDM-QPSK signal requires the BE to have 7 taps to obtain a negligible OSNR penalty.

I then evaluate the BE complexity for PDM-16QAM. As shown in Figure 4.4, a 7-tap BE is sufficient for NRZ PDM-16QAM, the same as that for PDM-QPSK. For a Nyquist PDM-16QAM signal with RRC 0.05 spectral shaping, the required BE tap numbers can be reduced from 15 to 3 if a matched filter is included in the CD equalizer, which is the same as that for PDM-QPSK, as expected. Note that for Nyquist 16QAM, slightly better performance can be obtained when a matched filter is used. Unlike PDM-QPSK, for the Nyquist-16QAM signal, the convergence speed can be significantly improved when the matched filter is included in the CD equalizer, as shown in Figure 4.5. This is due to the fact that the additional RDE algorithm is required for PDM-16QAM.

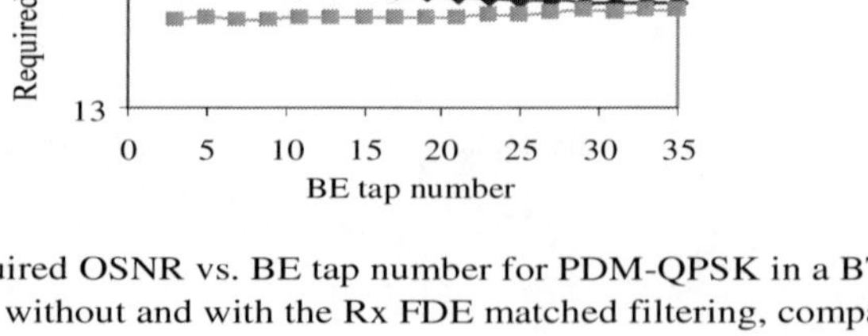

Figure 4.2. Required OSNR vs. BE tap number for PDM-QPSK in a BTB system, using RRC 0.05 without and with the Rx FDE matched filtering, compared to NRZ.

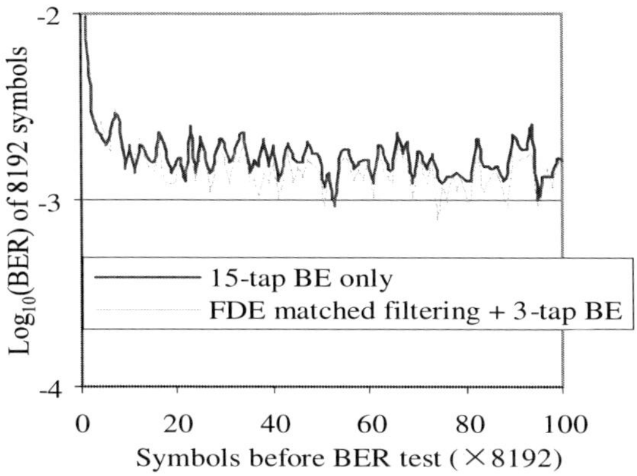

Figure 4.3. BER vs. Symbols before BER test for Nyquist PDM-QPSK in a BTB system, without and with the Rx FDE matched filtering (BE algorithm CMA, OSNR 13 dB.

Figure 4.3 plots the convergence speed of the BE in the two cases at OSNR of 13 dB. It shows that the convergence speeds of the BE without and with including the matched filter in the CD equalizer are almost the same. I believe this is due to the fast convergence property of CMA on PDM-QPSK.

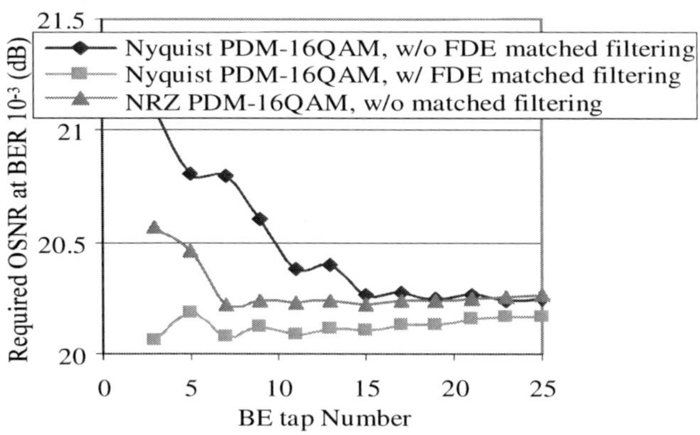

Figure 4.4. Required OSNR vs. BE tap number for PDM-16QAM in a BTB system, using RRC 0.05 without and with the Rx FDE matched filtering, compared to NRZ.

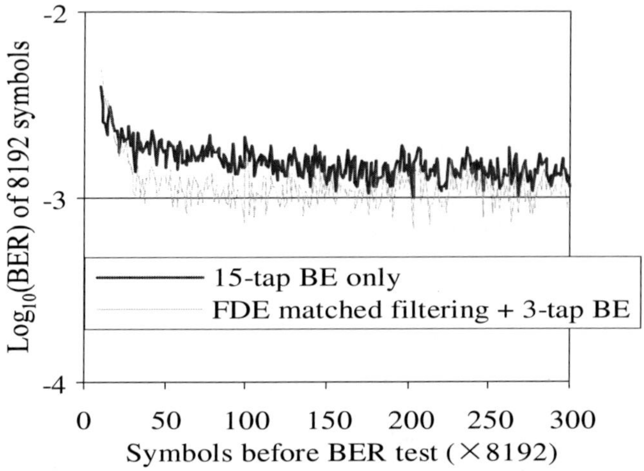

Figure 4.5. BER vs. Symbols before BER test for Nyquist PDM-16QAM in a BTB system, without and with the Rx FDE matched filtering (10×8192 symbols for CMA then switched to RDE, OSNR 20 dB).

One may wonder if adding the matched filtering function to the CD equalizer increases its complexity. To investigate this problem, we add 1000 km SSMF in the transmission link and use an overlap-and-save FDE with a size of 1024 to compensate for the CD. In general, FDE complexity can be represented by its overlap length, with a longer overlap length meaning higher complexity [37]. Figure 4.6 shows the overall FDE overlap length vs. required OSNR for a Nyquist PDM-QPSK signal with RRC 0.05 spectral shaping without and with including a matched filter in the CD equalizer. The BE tap numbers for the cases without and with a matched filter are 15 and 3, respectively. I also plot the result of a NRZ signal with a 7-tap BE in the figure for reference. The required FDE overlap length to achieve <0.1 dB extra OSNR penalty is about the same (~192 points) for the two cases i.e., adding a matched filter in the CD equalizer does not add any extra complexity to the equalizer. One reason for this is the matched filter and the CD filter are orthogonal to each other in the sense that one is all about the amplitude and the other is all about the phase in the frequency domain. Another reason is the matched filter requires a much shorter overlap length than the CD filter. Note that the required FDE overlap length for the NRZ system is about 224 points, which is because that the NRZ signal has a wider bandwidth than the Nyquist signal.

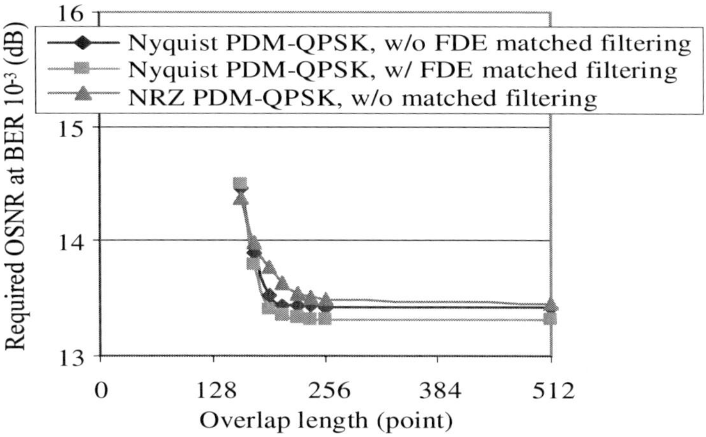

Figure 4.6. Required OSNR vs. FDE Overlap length of the CD equalizer for PDM-QPSK after 1000 km SSMF, without and with the Rx matched filtering.

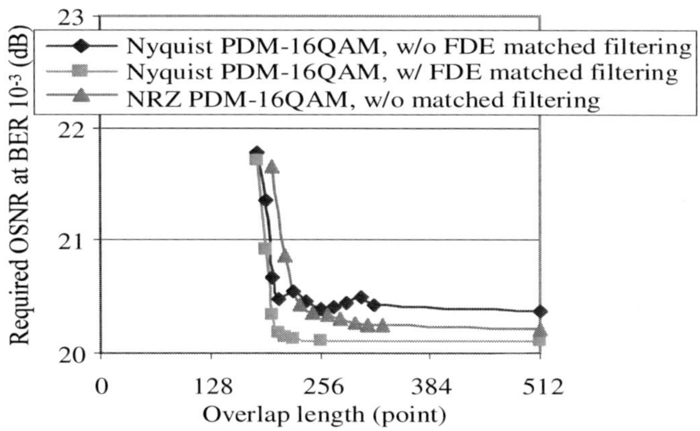

Figure 4.7. Required OSNR vs. FDE Overlap length of the CD equalizer for PDM-16QAM after 1000 km SSMF, without and with the Rx matched filtering.

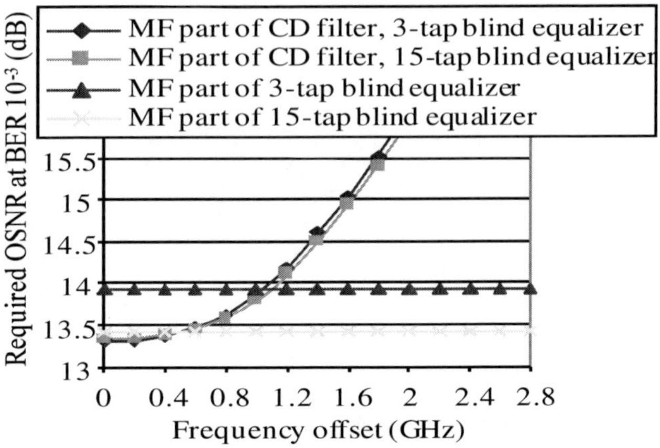

Figure 4.8. Required OSNR vs. frequency offset for PDM-QPSK with RRC 0.05.

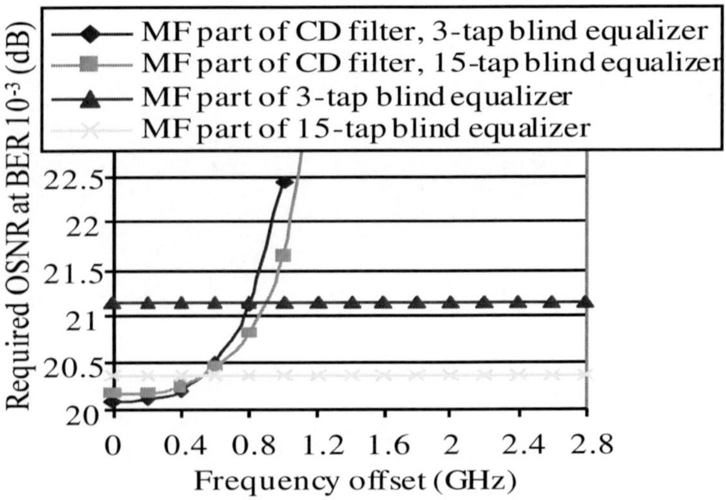

Figure 4.9. Required OSNR vs. frequency offset for PDM-16QAM with RRC 0.05.

Figure 4.7 shows the similar result for the PDM-16QAM system. For the Nyquist-16QAM signal with RRC 0.05 spectral shaping, the required overall

FDE overlap length to achieve <0.1 dB extra OSNR penalty is the same (~208 points) regardless of if the matched filter is included in the CD equalizer or not. For a regular NRZ PDM-16QAM signal, the required FDE overlap length is about 280 points, larger than that for the Nyquist 16QAM signal.

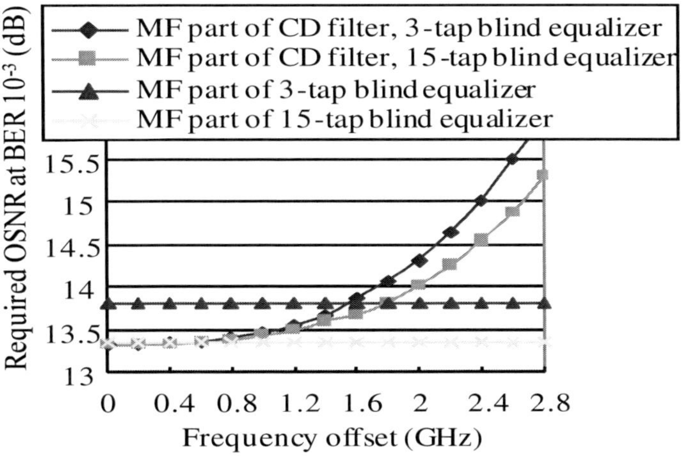

Figure 4.10. Required OSNR vs. frequency offset for PDM-QPSK with RRC 0.1.

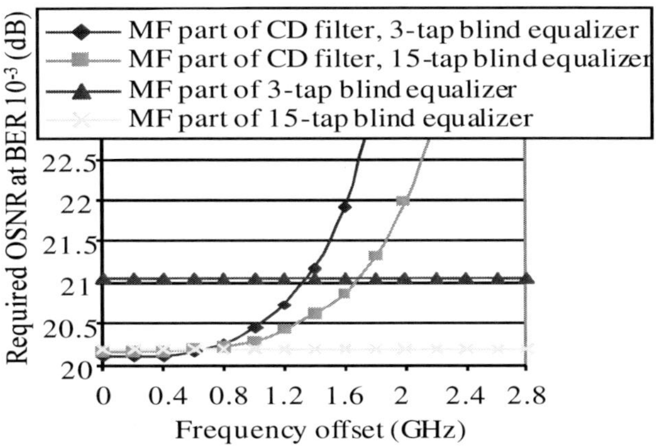

Figure 4.11. Required OSNR vs. frequency offset for PDM-16QAM with RRC 0.1.

We next investigate the impact of the frequency offset between transmitter and receiver, when the MF is included in the CD equalizer. The frequency offset causes a frequency shift of the baseband signal obtained by the coherent receiver. This frequency shift causes a position mismatch between a static MF and the baseband signal. We use Figs. 4.8-4.11 to show the penalty from the frequency offset. In the simulation, 17000-ps/nm CD is added, the FDE FFT size is 1024 at an oversampling of 2, and the overlap length of the FDE is 512 points. Fig. 4.8 and Fig. 4.9 show that, with a roll-off factor of 0.05, the OSNR penalty at BER 10^{-3} from 1 GHz frequency offset is about 0.5 dB for 28-Gbaud PDM-QPSK, and about 1.5 dB for 28-Gbaud PDM-16QAM. A larger roll-off factor of RRC spectral shaping reduces the penalty from the frequency offset, as shown in Fig. 4.10 and Fig. 4.11. Fig. 4.10 and Fig. 4.11 show that, with a roll-off factor of 0.1, the OSNR penalty at BER 10^{-3} from 1 GHz frequency offset is less than 0.2 dB for 28-Gbaud PDM-QPSK, and less than 0.4 dB for 28-Gbaud PDM-16QAM. The reason why a large roll-off factor reduces the frequency offset induced penalty is because the roll-off of the spectrum is slower, which reduces the portion of the signal that is cut off by the MF due to the frequency offset. We expect that with some monitoring for the frequency offset, the frequency offset can be reduced to less than 1 GHz and its impact on a receiver with a MF implemented as part of the CD filter will be negligible.

CONCLUSION

Nyquist-WDM is a very practical technique to increase the fiber capacity. Intensive researches on this topic have been done. From these researches, there are the following conclusions.

The complexity of the required transmit -side DSP is investigated to generate Nyquist-WDM QPSK and 16QAM signals by simulations. For a RRC 0.1 PDM-QPSK signals a FIR filter need 25 taps, while for a RRC 0.1 PDM-16QAM signal a FIR need 37 taps. A FDE filter with a FFT size of 64 and an overlap length of 8 points gives a slightly better performance than a 17-tap FIR filter, for RRC 0.1 PDM-QPSK signals. There are feasible structures to optimize the hardware complexity of FIR filters for Nyquist spectral shaping, so FIR filters are usually used for Nyquist spectral shaping.

The effects of DAC sampling speeds on the performance of Nyquist-WDM systems are also studied. The electrical filter shape and spectral pre-emphasis have a big impact on the system performance. With a 4th-order super-Gaussian electrical filter and spectral pre-emphasis, 1.2-samples/symbol DACs can generate 16QAM Nyquist-WDM signals at 1.05-symbol-rate channel spacing, with an OSNR penalty below 0.2 dB than 2-samples/symbol DACs.

The required spectral pre-emphasis, bandwidth and DAC resolution are studied for a 2-channel interleaved DAC using zero-order holding at the symbol rate to generate Nyquist spectrally shaped signals. It's also shown how a polyphase FIR filter structure can be applied in such transmitter DSP, to reduce the required multiplications by half and to increase the processing parallelism. With proper spectral pre-emphasis, a 2-channel interleaved DAC can generate Nyquist spectrally shaped signals with a negligible required OSNR penalty (below 0.1 dB). Both the required E-filter bandwidth and DAC resolution are similar to those with a conventional DAC used. To have a required OSNR penalty below 0.1 dB for a RRC 0.1 Nyquist-WDM system with a spectral shaping FIR filter tap number of 37, a DAC need a Gaussian E-filter bandwidth between 0.3 and 0.8 times the symbol rate for PDM-QPSK Nyquist systems, and between 0.35 and 0.65 times the symbol rate for PDM-16QAM Nyquist systems. And the required DAC resolution is 6 bits for PDM-QPSK Nyquist systems, and 7 bits for PMD-16QAM Nyquist systems.

We numerically investigated the required blind equalizer complexity in PDM-QPSK and PDM-16QAM Nyquist systems with RRC spectral shaping. We showed that including the matched filtering function in the CD equalizer can significantly reduce the complexity of the blind equalizer, without adding additional complexity to the CD equalizer. We also show that the laser frequency offset between the local oscillator at the receiver and the laser at the transmitter causes penalties to the system when a static matched filter is used in the receiver. A larger roll-off factor of RRC spectral shaping reduces the penalty from the frequency offset. With a roll-off factor of 0.1, the OSNR penalty at a BER of 10^{-3} from 1 GHz frequency offset is less than 0.2 dB for 28-Gbaud PDM-QPSK, and less than 0.4 dB for 28-Gbaud PDM-16QAM.

REFERENCES

[1] Shieh, W; Bao, H; Tang, Y."Coherent optical OFDM: theory and design," *Opt. Exp.*, vol. 16, no. 2, 841-859, 2008.

[2] Liu, X; Chandrasekhar, S; Zhu, B; Winzer, PJ; Gnauck, AH; Peckham, DW."Transmission of a 448-Gb/s reduced-guard-interval CO-OFDM signal with a 60-GHz optical bandwidth over 2000 km of ULAF and five

80-GHz-Grid ROADMs," in *Proc. Opt. Fiber Commun./Nat. Fiber Opt. Eng. Conf.*, 2010, paper PDPC2, 1-3.

[3] Cai, J.-X; Cai, Y; Davidson, CR; Foursa, DG; Lucero, A; Sinkin, O; Patterson, W; Pilipetskii, A; Mohs, G; Bergano, NS."Transmission of 96×100G pre-filtered PDM-RZ-QPSK channels with 300% spectral efficiency over 10,608km and 400% spectral efficiency over 4,368km," in *Proc. Opt. Fiber Commun./Nat. Fiber Opt. Eng. Conf.*, 2010, paper PDPB10, 1-3.

[4] Bosco, G; Carena, A; Curri, V; Poggiolini, P; Torrengo, E; Forghieri, F."Investigation on the robustness of a Nyquist-WDM Terabit super channel to transmitter and receiver non-idealities," in *Proc. Eur. Conf. Opt. Commun.*, 2010, paper Tu.3.A.4, 1-3.

[5] Salsi, M; Koebele, C; Tran, P; Mardoyan, H; Dutisseuil, E; Renaudier, J; Bigot-Astruc, M; Provost, L; Richard, S; Sillard, P; Bigo, S; Charlet, G."Transmission of 96×100Gb/s with 23% super-FEC Overhead over 11,680 km, using Optical Spectral Engineering," in *Proc. Opt. Fiber Commun./Nat. Fiber Opt. Eng. Conf.*, 2011, paper OMR2, 1-3.

[6] Xie, C; Raybon, G; Winzer, PJ."Hybrid 224-Gb/s and 112-Gb/s PDM-QPSK transmission at 50-GHz channel spacing over 1200-km dispersion-managed LEAF® spans and 3 ROADMs," in *Proc. Opt. Fiber Commun./Nat. Fiber Opt. Eng. Conf.*, 2011, paper PDPD2, 1-3.

[7] Igarashi, K; Mori, Y; Katoh, K; Kikuchi, K."Bit-error Rate Performance of Nyquist Wavelength-Division Multiplexed Quadrature Phase-Shift Keying Optical Signals," in *Proc. Opt. Fiber Commun./Nat. Fiber Opt. Eng. Conf.*, 2011, paper OMR6, 1-3.

[8] Zhou, X; Nelson, L; Magill, P; Isaac, R; Zhu, B; Peckham, DW; Borel, P; Carlson, K."8 450-Gb/s, 50-GHz-spaced, PDM-32QAM transmission over 400km and one 50GHz-grid ROADM," in *Proc. Opt. Fiber Commun./Nat. Fiber Opt. Eng.* Conf., 2011, paper PDPB3, 1-3.

[9] Chen, S; Xie, C; Zhang, J"Comparison of Advanced Detection Techniques for QPSK Signals in Super-Nyquist WDM Systems", IEEE Photonics Technology Letters, Oct. 2014.

[10] Morita, Iand Peng, W.-R."High capacity WDM transmission using terabit super-channels", European Conference and Exhibition on Optical Communication (ECOC 2012), Super Channel Transmission (Th.2.C), Amsterdam, Netherlands, Sep. 2012.

[11] Zhang, J; Yu, J; Chi, N"Generation and transmission of 512-Gb/s quad-carrier digital super-Nyquist spectral shaped signal", Optics Express, Vol. 21, Issue 25, 31212-31217 (2013).

[12] Châtelain, B; Laperle, C; Roberts, K; Xu, X; Chagnon, M; Borowiec, A; Gagnon, F; Cartledge, J; Plant, DV."Optimized pulse shaping for intra-channel nonlinearities mitigation in a 10 Gbaud dual-polarization 16-QAM system," in *Proc. Opt. Fiber Commun./Nat. Fiber Opt. Eng. Conf.*, 2011, paper OWO5, 1-3.

[13] Savory, SJ."Digital filters for coherent optical receivers," *Opt. Exp.*, vol. 16, no. 2, 804-817, 2008.

[14] Falconer, D; Ariyavisitakul, SL; Benyamin-Seeyar, A; Eidson, B."Frequency Domain Equalization for Single-Carrier Broadband Wireless Systems," *IEEE Commun. Mag.*, vol.40, no 4, 58-66, 2002.

[15] Pancaldi, F; Vitetta, GM; Kalbasi, R; AI-Dhahir, N; Uysal, M; Mheidat, H."Single-carrier frequency domain equalization," *IEEE Signal Process. Mag.*, vol. 25, no. 5, 37-56, 2008.

[16] Md. Faruk, S; Kikuchi, K."Adaptive frequency-domain equalization in digital coherent optical receivers," *Opt. Exp.*, vol. 19, no. 13, 12789-12798, 2011.

[17] Kudo, R; Kobayashi, T; Ishihara, K; Takatori, Y; Sano, A; Miyamoto, Y."Coherent Optical Single Carrier Transmission Using Overlap Frequency Domain Equalization for Long-Haul Optical Systems," *IEEE J. Lightw. Technol.*, vol. 27, no. 16, 3721-3728, 2009.

[18] Geyer, J.C; Fludger, C.R.S; Duthel, T; Schulien, C; Schaauss, B."Efficient frequency domain chromatic dispersion compensation in a coherent Polmux QPSK-receiver," in *Proc. Opt. Fiber Commun./Nat. Fiber Opt. Eng. Conf.*, 2010, paper OWV5, 1-3.

[19] Chen, C; Zhuge, Q; Plant, DV."Reduced-Guard-Interval CO-OFDM with Overlapped Frequency-Domain CD and PMD Equalization," in *Proc. Opt. Fiber Commun./Nat. Fiber Opt. Eng. Conf.*, 2011, paper OWE7, 1-3.

[20] Spinnler, B."Equalizer Design and Complexity for Digital Coherent Receivers," *IEEE J. of Sel. Topic. in Quant. Electr.*, 2010, Vol. 16, No. 5, pages: 1180-1192.

[21] Proakis, JG. *Digital Communications*, 4th ed., chap. 9, New York, McGGraw-Hill, 2001.

[22] Wang, J; Xie, C; Pan, Z."Generation of Spectrally Efficient Nyquist-WDM QPSK Signals using DSP Techniques at Transmitter," in *Proc. Opt. Fiber Commun./Nat. Fiber Opt. Eng. Conf.*, 2012, Paper OM3H.5, 1-3.

[23] Wang, J; Xie, C; Pan, Z."Generation of Spectrally Efficient Nyquist-WDM QPSK Signals using digital FIR or FDE filters at Transmitters," *IEEE J. Lightw. Technol.*, vol. 30, issue 23, pages:3679-3686, 2012.
[24] Wang, H; Chan, C; Choy, C."High Speed CMOS Digital-to-Analog Converter With Linear Interpolator," *IEEE Trans. on Cons. Electr.*, vol. 46, 1137-1142, 2000.
[25] Winzer, PJ; Pfennigbauer, M; Essiambre, R.-J."Coherent Crosstalk in Ultradense WDM Systems," *IEEE J. Lightw. Technol.*, vol. 23, no. 4, 1734-1744, 2005.
[26] Wang, J; Xie, C; Pan, Z."Optimization of DSP to Generate Spectrally Efficient 16QAM Nyquist-WDM Signals," *IEEE Photon. Technol. Lett.*, vol. 25, no. 8, 772-775, 2013
[27] Zhu, W.-P; Ahmad, MO; Swamy, M.N.S."ASIC implementation architecture for pulse shaping FIR filters in 3G mobile communications," in *Proc. ISCAS*, 2002, vol. 1, I-433 - I-436.
[28] Cartledge, JC; Downie, JD; Hurley, JE; Karar, AS; Jiang, Y; Roberts, K."Pulse shaping for 112 Gbit/s polarization multiplexed 16-QAM signals using a 21 Gsa/s DAC," *Opt. Exp.*, vol. 19, no. 26, B628-B635, 2011.
[29] Duthel, T; Raabe, C; Hermann, P; Whiteaway, J.E.A; Geyer, J.C; Fludger, C.R.S, and Kupfer, T."DAC Enabled Spectrally Efficient CP-QPSK at 28 Gbaud," in *Proc. Eur. Conf. Opt. Commun.*, 2012, paper Tu.4.A.4, 1-3.
[30] Shannon, CE."Communication in the presence of noise," in *Proc. Institute of Radio Engineers*, vol. 37, no. 1, 10–21, Jan. 1949. Reprint as classic paper in *Proc. IEEE*, vol. 86, no. 2, 1998.
[31] Sklar, B. *Digital Communications*, 2nd edition, Prentice-Hall 2001, 69.
[32] Fatadin, I; Ives, D; Savory, SJ."Blind Equalization and Carrier Phase Recovery in a 16-QAM Optical Coherent System," *IEEE J. Lightw. Technol.*, vol. 27, no. 15, 3042-3049, 2009.
[33] Randel, S; Pilori, D; Corteselli, S; Raybon, G; Adamiecki, A; Gnauck, A; Chandrasekhar, S; Winzer, P; Altenhain, L; Bielik, A; Schmid, R."All-Electronic Flexibly Programmable 864-Gb/s Single-Carrier PMD-64-QAM"," in *Proc. OFC/NFOEC, 2014*, paper Th5C.8
[34] Han, C; Igarashi, K; Kikuchi, K."Influence of Channel Misalignment of Time-interleaved DAC on Sensitivity Degradation in Coherent Optical Receivers," in *Proc. OFC/NFOEC, 2013*, paper OTh1F
[35] Wang, J; Xie, C; Pan, Z."Reducing Equalizer Complexity in Coherent Receivers for Nyquist Spectrally Shaped Systems with Matched Filters,"

in *Proc. Opt. Fiber Commun./Nat. Fiber Opt. Eng. Conf.*, 2013, paper OTu2I.3, 1-3.

[36] Wang, J; Xie, C; Pan, Z."Matched Filter Design for RRC Spectrally Shaped Nyquist-WDM Systems," *IEEE Photon. Technol. Lett.*, vol. 25, no. 23, 2263-2266, 2013

[37] Leibrich, J; Rosenkranz, W." Frequency Domain Equalization with Minimum Complexity in Coherent Optical Transmission Systems," in *Proc. Opt. Fiber Commun./Nat. Fiber Opt. Eng. Conf.*, 2010, paper OWV1, 1-3.

In: Optical Communication Systems
Editor: Karla P. Boyd
ISBN: 978-1-63482-833-8

Chapter 4

YAGI-UDA NANOANTENNA IN THE NEAR-FIELD OPTICAL DOMAIN

***M. R. K. Soltanian*[*1], *I. S. Amiri*[*,1,2]**
***S. D. Emami*[1] *and H. Ahmad*[1]**
[1] Photonics Research Centre, University of Malaya,
Kuala Lumpur, Malaysia
[2] Laser Center, Ibnu Sina ISIR,
Universiti Teknologi Malaysia (UTM),
Johor Bahru, Malaysia

ABSTRACT

An optical analogue of the well-established radio-frequency Yagi-Uda antenna, stands out by its effective unidirectional light emission and enhancement. Similar to radio wave and microwave antennas, their purpose is to convert the energy of free propagating radiation to localized energy, and vice versa. Optical antennas exploit the unique properties of metal nanostructures, which behave as strongly coupled plasmas at optical frequencies. Nobel metallic nanoparticles like Gold or silver with their plasmon resonances in the optical wavelength region can be used in a similar manner. Near-field coupling of quantum emitters to these particle plasmon resonances is used to redirect the light emission. Pursuing a brief introduction to the emerging field of optical

[*] Email Address: amiri@um.edu.my and soltanian79@gmail.com

nanoantennas, here we review the effect of changing the dipole, director and reflector's radii on the directivity of Yagi-Uda nanoantenna in near-field optical domain. The various aspects of the guiding properties of arrays of metamaterial or plasmonic particles, with a specific focus on Yagi-Uda nanoparticle models are investigated. In-depth analysis is performed on the transmission, reflection, and directive properties of nanoantenna.

Keywords: optical nanoantennas, Yagi-Uda antennas, plasmonics, nanoparticles

1. INTRODUCTION

Plasmonic nanoparticles are known as key elements in the control of light interaction with quantum emitters. In interaction with light, noble metal nanostructures, localize nanometer light wavelength in tiny volumes and enhance the radiative decay rates of nearby emitters. This is due to smaller features of nanoparticles than the wavelength of the excitation light, which can excite evanescent waves that cover a wide frequency range. Higher ability to control the angular emission of single molecules brings about more attention to plasmonic nanoantennas recently. High directivity facilitates of quantum emitter excitation by a collimated beam and the collection of the radiated light is the other key features of optical nanoantennas. Among the plasmonic nanoantennas, Yagi-Uda has been successfully introduced for higher angular directivity at optical frequencies [1-9]. It is easy to construct, and it has a high efficiency in terms of directivity [10] and losses that happens among the antenna as it is operating at a given frequency. Moreover, it can be shrunk to become small as 1/25 of a human hair with minimum changes in construction to operate in the infrared and visible regions. A generic Yagi-Uda antenna comprises of an actively driven element which is called feed surrounded by passive elements, which are not driven [Fig. 1(a)]. They are two types: first, a resonant reflector and second, a few equally separated directors. The key parameter of the whole structure is the length of the feeding element resonating the desired frequency. The lengths of the reflector and directors are preferred to be off resonance and designed to add up in phase in the forward direction, and to cancel in the reverse direction. For that purpose, the director elements are made shorter and the reflector is made longer than the feeding element, which leads to in– and out– of-phase current distributions.

The Yagi-Uda nanoantenna consists of dipole, director, and a reflector. Single emitters such as atoms and molecules are inherently quantum objects with two levels energy system can be represented by a classical dipole. Director elements generally consisting of a finite chain of identical particles that helps to transmit the energy in a particular direction and a reflector element typically based on a slightly larger particle [11-13]. Nearfield coupling of dipole to Yagi-Uda nanoantenna lead to light emission redirection, and the farfield radiation is completely determined by the antenna mode. The redirection of emission is defined by directivity. Directivity is measured in decibels of the intensity along the collector axis divided by the steradian average of total radiated power [14].

2. RESULTS AND DISCUSSION

In this research work, the effect of dipole displacement and reflector's radius on the Yagi-Uda nanoantenna has been examined carefully. Fig. 1.a shows the Yagi-uda nanoantenna directivities at different reflector's radiuses. The director's radiuses are fixed at 70 nm. Refer to Fig. 1.a the optimized reflector radius is achieved at 95 nm reflector's radiuses. It is widely regarded that the reflecting/collecting nature of the particle depends solely on the polarizability of the nanoparticle: i.e., when the frequency of the emitter is larger than a particle's plasmon frequency, the particle will act as a reflector and when it is lower, the particle will act as a collector [15]. This is extremely depends on the phase shift that the polarizability of the nanoparticles has to dominant. Fig. 1.b and Fig 1.c show the E-field pattern in λ=618 nm and λ=800 nm respectively. Refer to figures it is clearly illustrates strong reflection from reflector and the radiation enhancement at the collectors direction in directivity peak around λ=618 nm rather than λ=800 nm.

Determining the optimum directivity of the Yagi-Uda nanoantenna in a required attention on three variables, these being nanoparticle radius, distance between the dipole resonator and nano-particles, and gap between each nanoparticle in the model structure as shown in Fig. 2. Fig. 2(A) illustrates the variations of directivity of the nanoantenna when the nanoparticle radii was varied from 50 to 100 nm and gap between two adjacent nanoparticles changed from 120 to 220 nm, while the distance from the dipole remained constant at 200 nm.

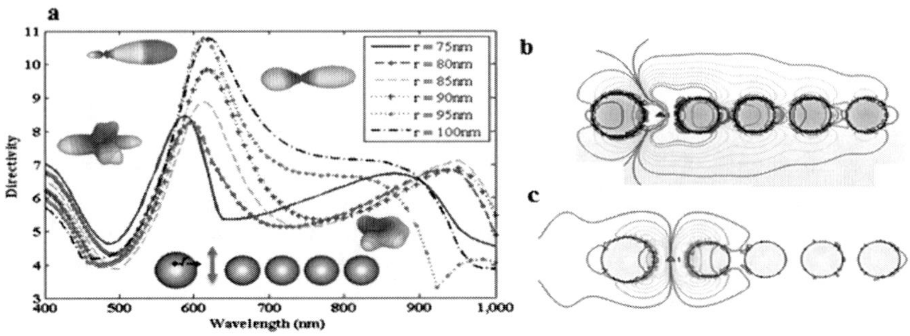

Figure 1. a) directivity behavior of Yagi-Uda nanoantennas in term of different reflector's radius and emitter wavelength. b) The E-field pattern in λ=618 nm. c) The E-field pattern in λ=800 nm.

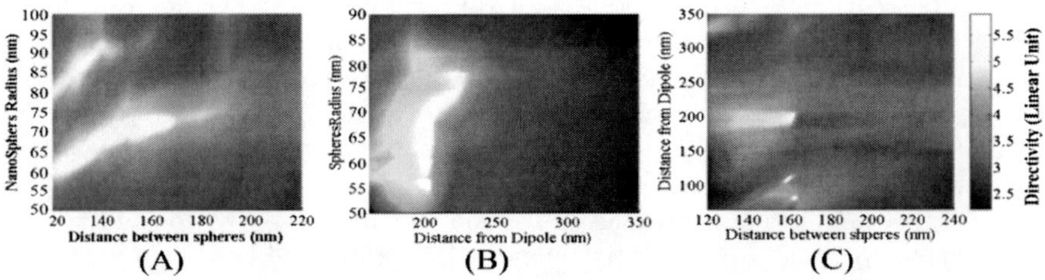

Figure 2. Directivity of the nanoantenna (A) nanosphere radius (nm) versus space between nanospheres (nm) (B) nanosphere radius (nm) versus distance form dipole (nm) (C) distance from dipole (nm) versus space between nanospheres (nm).

The highest directivity could be observed at 5.8 dBi, which corresponded to a distance of 160 nm between nanoparticles of 70 nm radii. As the pentamers moved apart from each other, the coupling factor between the nanospheres decreased drastically. Fig. 2(B) shows the effects of varying nanoparticle radius when the Yagi-Uda nanoantenna was repositioned away from the dipole and gap distance between the nanoparticles remained at 20 nm. Observation of Fig. 2(B) confirms that directivity changes were rather subtle when nanoparticle size was varied within a 50 to 75 nm range. Fig. 2 (C) shows the change of directivity mode for distance between the nanospheres when varied from 120 to 240 nm and for the nanoantenna when moved away from the dipole by 100 to 350 nm. The changes of nanoparticles radii within a reasonable range were not as effective in comparison to variation of gap between the nanoparticles. The most effective dipole distance from the nanoparticles was measured at 200 nm while the optimum size was 70 nm and distance of nanoparticles in a state of subradiant directivity was 160 nm. Based on Fig. 2(B) the highest directivity achieved when the distance from dipole is about 200 nm and this directivity is remains almost constant by varying the spheres' radius from 55 to 75 nm.

Conclusion

This paper described a detailed analysis of the various aspects of the guiding properties of arrays of metamaterial or plasmonic particles, with a specific focus on Yagi-Uda nanoparticle models. In-depth analysis was performed on the transmission, reflection, and directive properties of nanoantenna. The performance properties of analytically-designed nanoantenna were numerically simulated using CST microwave studio and MATLAB. Results for the proposed model showed directivity of the nanoantenna enhanced by increasing the radius of reflector.

References

[1] Dregely, D., Taubert, R., Dorfmuller, J., Vogelgesang, R., Kern, K. and Giessen, H. 3D optical Yagi-Uda nanoantenna array. *Nat Commun*, 2(04/05/online 2011), 267.

[2] Maksymov, I. S., Staude, I., Miroshnichenko, A. E. and Kivshar, Y. S. Optical yagi-uda nanoantennas. *Nanophotonics*, 1, 1 2012), 65-81.

[3] Kosako, T., Kadoya, Y. and Hofmann, H. F. Directional control of light by a nano-optical Yagi–Uda antenna. *Nature Photonics*, 4, 5 2010), 312-315.

[4] Dregely, D., Taubert, R., Dorfmüller, J., Vogelgesang, R., Kern, K. and Giessen, H. 3D optical Yagi-Uda nanoantenna array. *Nature communications*, 22011), 267.

[5] Li, J., Salandrino, A. and Engheta, N. Shaping light beams in the nanometer scale: A Yagi-Uda nanoantenna in the optical domain. *Physical Review B*, 76, 24 2007), 245403.

[6] Li, J., Salandrino, A. and Engheta, N. Optical spectrometer at the nanoscale using optical Yagi-Uda nanoantennas. *Physical Review B*, 79, 19 2009), 195104.

[7] Krasnok, A. E., Miroshnichenko, A. E., Belov, P. A. and Kivshar, Y. S. Huygens optical elements and Yagi—Uda nanoantennas based on dielectric nanoparticles. *JETP letters*, 94, 8 2011), 593-598.

[8] Dorfmuller, J., Dregely, D., Esslinger, M., Khunsin, W., Vogelgesang, R., Kern, K. and Giessen, H. Near-field dynamics of optical Yagi-Uda nanoantennas. *Nano letters*, 11, 7 2011), 2819-2824.

[9] Taminiau, T. H., Stefani, F. D. and van Hulst, N. F. Enhanced directional excitation and emission of single emitters by a nano-optical Yagi-Uda antenna. *Optics express*, 16, 14 2008), 10858-10866.

[10] Balanis, C. A. *Antenna theory: analysis and design*. John Wiley & Sons, 2012.

[11] Bonod, N., Devilez, A., Rolly, B., Bidault, S. and Stout, B. Ultracompact and unidirectional metallic antennas. *Physical Review B*, 82, 11 2010).

[12] Stout, B., Devilez, A., Rolly, B. and Bonod, N. Multipole methods for nanoantennas design: applications to Yagi-Uda configurations. *JOSA B*, 28, 5 2011), 1213-1223.

[13] Devilez, A., Stout, B. and Bonod, N. Compact metallo-dielectric optical antenna for ultra directional and enhanced radiative emission. *ACS nano*, 4, 6 2010), 3390-3396.

[14] B. Stout, A. D., B. Rolly, N. Bonod Multipole methods for Nano-antennas design: applications to Yagi-Uda configurations. *J. Opt. Soc. Am. B*, 1213-1223 2011).

[15] Rolly, B., Stout, B., Bidault, S. and Bonod, N. Crucial role of the emitter?particle distance on the directivity of optical antennas. *Opt. Lett.*, 36, 17 2011), 3368-3370.

INDEX

O

P

Q

R

S